Lecture Notes in Mathematics

Edited by A. Dold and B. Eckmann

Subseries: Nankai Institute of Mathematics, Tianjin, P.R. China
vol. 2
Adviser: S.S. Chern

1241

Lars Gårding

Singularities in Linear Wave Propagation

Springer-Verlag
Berlin Heidelberg New York London Paris Tokyo

Author

Lars Gårding
Department of Mathematics, University of Lund
Box 118, 22100 Lund, Sweden

Mathematics Subject Classification (1980): 35-xx; 35Lxx

ISBN 3-540-18001-X Springer-Verlag Berlin Heidelberg New York
ISBN 0-387-18001-X Springer-Verlag New York Berlin Heidelberg

Library of Congress Cataloging-in-Publication Data. Gårding, Lars, 1919- Singularities in linear wave propagation. (Lecture notes in mathematics; 1241) Includes index. 1. Differential equations, Hyperbolic. 2. Wave motion. Theory of. 3. Singularities (Mathematics) I. Title. II. Series: Lecture notes in mathematics (Springer-Verlag); 1241. QA3.L28 no. 1241 510 s 87-16397 [QA377] [515.3'53]
ISBN 0-387-18001-X (U.S.)

Printing and binding: Druckhaus Beltz, Hemsbach/Bergstr.
2146/3140-543210

TABLE OF CONTENTS

Historical introduction 1
Chapter 1 Hyperbolic operators with constant coefficients 10
1.1 Algebraic hyperbolicity 10
1.2 Distributions associated with the inverses of a homogeneous hyperbolic polynomial 15
1.3 Intrinsic hyperbolicity 17
1.4 Fundamental solutions of homogeneous hyperbolic operators. Propagation cones. Localization. General conical refraction 20
1.5 The Herglotz-Petrovsky formula 25
Chapter 2 Wave front sets and oscillatory integrals 34
2.1 Wave front sets 34
2.2 The regularity function 37
2.3 Oscillatory integrals 39
2.4 Fourier integral operators 44
2.5 Applications 45
Chapter 3 Pseudodifferential operators 51
3.1 The calculus of pseudodifferential operators 53
3.2 L^2 estimates. Regularity properties of solutions of pseudodifferential equations 62
3.3 Lax's construction for Cauchy's problem and a first order differential operator 70
Chapter 4 The Hamilton-Jacobi equation and symplectic geometry 73
4.1 Hamilton systems 73
4.2 Symplectic spaces and Lagrangian planes 75
4.3 Lagrangian submanifolds of the cotangent bundle of a manifold 78
4.4 Hamilton flows on the cotangent bundle. Very regular phase functions 81
Chapter 5 A global parametrix for the fundamental solution of a first order hyperbolic pseudodifferential operator 85
5.1 Cauchy's problem for a first order hyperbolic pseudodifferential operator 85
5.2 Cauchy's problem on the product of a line and a manifold 88
5.3 A global parametrix 89
Chapter 6 Changes of variables and duality for general oscillatory integrals 95
6.1 Hörmander's equivalence theorem for oscillatory integrals with a regular phase function 96
6.2 Reduction of the number of variables 98
6.3 Duality and reduction of the number of variables 101
Chapter 7 Sharp and diffuse fronts of paired oscillatory integrals 104
7.1 A family of distributions 105
7.2 Polar coordinates in oscillatory integrals 108
7.3 Almost analytic extensions 111
7.4 Singularities of paired oscillatory integrals with Hessians of corank 2 112
7.5 The general case. Petrovsky chains and cycles. The Petrovsky condition 121
References 124
Index 125

SINGULARITIES IN LINEAR WAVE PROPAGATION

Lars Gårding

Historical introduction The theory of wave propagation started with Huyghens's theory of wave front sets as envelopes of elementary waves. Its first success was the proper explanation of the propagation of light in refracting media. Its modern successor is the theory of boundary problems for hyperbolic systems of partial differential equations. The development which lead to this theory is a story of a search for proper mathematical tools.

The first chapter is the discovery in the eighteenth century of a paradox. The wave equation $u_{tt}-u_{xx}=0$ in one time and one space dimension expresses the movement of the deviation u from rest position for an idealized string. Its general solution $f(x-t)+g(x+t)$ with f and g arbitrary is the sum of two travelling waves with opposite directions. But it was also possible to express the movements of a string fixed at its end points as an infinite sum of sine functions. This raised the question about the nature of functions and how a series with smooth terms could express arbitrary functions. The first efficient solution of this problem came two hundred years later with the theory of distributions.

The nineteenth century made important discoveries about wave propagation. Gemetrical optics was developed to great perfection by Hamilton. It is a theory of normals of wave fronts, in other words of rays rather than waves of light. It gave a very good idea of wave fronts or caustics but not a very clear idea about their intensity or the intensity of light outside the fronts. Other efforts centered around the wave equation, in our notation and with the propagation velocity normalized to 1,

$$u_{tt}-\Delta u = 0,$$

where Δ is Laplace's operator in n space variables $x=(x_1,...,x_n)$. The physically interesting case is of course n=3. In the beginning of the nineteenth century it was observed that the spherical waves $u=f(t-|x|)/|x|$ are solutions for arbitrary functions f and Poisson discovered the remarkable formula, in modern notation,

$$u(t,x) = (4\pi)^{-1} \int f(y)\delta(t-|x-y|)dy/|x-y|,$$

which solves Cauchy's problem $u_{tt}-\Delta u=0$, u=0, $u_t=f(x)$ for t=0. Its fit with geometrical optics was perfect, the support of the solution at time t is precisely the envelope of spheres with radius t and centers in the support of f. For almost a century, this cemented the idea that geometrical optics contains almost the whole story of wave propagation.

In modern language we can interpret Poisson's formula by saying that the distribution

(1) $E(t,x) = H(t)\ \delta(t-|x|)/4\pi|x|$, H(t)=1 when t>0 and 0 otherwise,

which solves the wave equation $E_{tt}-\Delta E=\delta(t)\delta(x)$ and vanishes when t<0 describes the forward emission of light from a point source. It can also be described as the forward fundamental solution of the wave equation in three space variables.

The most interesting problem of geometrical optics was the problem of double refraction observed and analyzed already by Huygens. A ray of light entering certain kinds of crystals is refracted into two rays whose directions vary with the direction of the incident ray relative to the crystal. According to Huygens's theory of refraction, this means that light in the crystal has two velocities, both direction dependent, which can be measured by the strength of the refraction.

Directions where the two velocities are the same are called optical axes. They appear as double points of the velocity surface which is obtained by taking velocity as distance to an origin along a variable ray. For crystals with one optical axis, Huygens found the velocity surface to be a sphere and an ellipsoid tangent to it.

Crystals with two optical axes remained a mystery until the the French physicist Fresnel found explicit velocity surfaces for them. They turned out to be algebraic of degree 4 depending on three constants varying with the nature of the crystal. The surfaces are symmetric around the origin with two sheets which come together at four double points on the optical axes.

Associated with the velocity surface there is the wave surface consisting of wave fronts at time t=1 emanating from an instantaneous point source of light in the crystal. According to Huygens, the wave surface is the envelope of the velocity surface. Fresnel guessed its analytical form. By a freak of nature, it is identical to the velocity surface with the three constants inverted and hence it has the same general form as the velocity surface. The computations were carried out by, among others, Hamilton. He added an important complement, observing that the tangent planes to the velocity surface through a double point form a circle on the outer sheet of the wave surface which bounds a circular disc covering the inlet to a double point. He predicted from this that an outside ray of light whose direction coincides with an optical axis ought to be broken into a cone of rays. This phenomenon, the conical refraction, was verified by experiment a short time later.

Hamilton made his discovery in the late 1820's. The following decades saw extensive activity with the aim of understanding the nature of light. The first attempts were based on analogy with elasticity theory and resulted among other things in the equations of Lame, a 3X3 hyperbolic system of second order differential equations in four variables, time and three space variables. These equations are identical with what one gets from Maxwell's equations for the electric field in a dielectricum when the magnetic field is elimininated.

To solve the initial value problem for Lame's system was a great challenge taken up by Sonya Kovalevskaya. She had a model to go by, Weierstrass's solution of the Cauchy problem for an analogous system associated with the product of two wave operators with different speeds of light. Led by geometrical optics, she assumed that light from a point source ought to propagate between the two sheets of the wave surface leaving no trace behind. The latter assumption is correct but she did not realize that there is light also between the outer sheet and its convex hull. Her formal calculations where she used the fact that the wave front surface can be parametrized by elliptic functions led her astray. The solution that she deduced is identically zero, a clear contradiction with the Cauchy-Kovalevskya theorem. Her mistake was pointed out a few years later by Volterra. He corrected the formulas but did not arrive at a solution of Cauchy's problem. Earlier, his faith in geometrical optics as a complete clue to wave propagation was shaken when he found that the analogue of the distribution (1) for two space variables is

$$H(t)\ H(t^2 - |x|^2)^{-1/2}/2\pi,$$

which describes propagation of light from an instantaneous point source in a medium with two space variables. Since this distribution does not vanish when $|x|<t$, there is an afterglow behind the wave front on the circle $|x|=t$. In lectures that he gave in Stockholm in 1906, Volterra pointed out that the analytical tools tried so far were not sufficient to treat Lame's equations for the double refraction. One of his listeners, a young mathematics student Nils Zeilon, took notice. His admired teacher Ivar Fredholm had constructed fundamental solutions of elliptic differential operators in three variables using abelian integrals. Zeilon continued his work for other types of equations but using another point of departure, namely the remark that if $P(\xi)$ is a polynomial in n variables, the integral

$$(2) \qquad E(x)=(2\pi)^{-n} \int \exp ix.\xi\ d\xi/P(\xi)$$

is, at least formally, a fundamental solution for the operator $P(D)$

where $D=\partial/i\partial x$. In fact,

$$P(D)E(x)=(2\pi)^{-n}\int \exp ix.\xi \, d\xi = \delta(x).$$

The problem is to make sense of the integral (2) which may diverge at 0, at infinity and at the zeros of P. Apart from this, the formal machinery works also when $P(\xi)$ is a square mateix whose elements are polynomials, in particular for the Lame system. Zeilon's mathod of avoiding singularities was to move the chain R^n of integration into C^n. This can be done in various ways. Zeilon's intuition led him right, but his arguments are shady, read by a critical eye. This also applies to his magnum opus, two long articles around 1920 on the problem of conical refraction. But his results are right. The support of the fundamental solution includes the space the outer sheet of the wave surface and its convex hull. This fact has to do with conical refraction, but the precise connection was not clarified until 1961 with a paper by Ludwig.

Zeilon's work did not get much attention. Some years later Herglotz constructed forward fundamental solutions of hyperbolic differential operators wih constant coefficients in any number of variables. For them, the velocity surface has m sheets corresponding to m different propagation velocities. He applied the Fourier transform to the space variables and arrived at very simple formulas covering also the Lame system. He showed that the wave surface in the general case is a system of criss-crossing surfaces of varying dimensions near which the fundamental solution may have a very complicated behavior. Outside the wave surface, i.e. outside its fastest front, the fundamental solution is zero, but it may also vanish in regions inside the fastest front as is does for propagation of light in free space. These regions, the lacunas, attracted the the interest of Petrovsky who published a fundamental paper about them in the forties where he tied the existence if lacunas to topological properties of plane sections of the complex velocity surface. His work was extended to the general case of degenerate velocity surfaces by Atiyah, Bott and Gårding in the early seventies.

Fundamental solutions E_n of the wave equation with an arbitrary number of space variables were constructed by Tedone already in 1889 in the form of solutions of the corresponding Cauchy problems. In terms of the function

$$d(t,x)=t^2-|x|^2,$$

their main properties can be described as follows, They vanish outside the forward light cone where $t\geq 0$, $d(t,x)\geq 0$. Inside the light cone E_n behaves like

$$\text{const } d(t,x)^{(n-3)/2}$$

when n is even. It vanishes there when n is odd >1 and behaves like

$$\text{const } \delta^{((n-3)/2)}(d(t,x))$$

on the light cone outside the origin. Tedone's results were extended to variable coefficients by Hadamard in a famous book, The Cauchy Problem and Hyperbolic Linear Partial Differential Equations, published in 1923 with a French edition in 1932. He replaced the wave operator by an operator

$$P=\Sigma\, g_{ik}(x)u_{ik} + \text{lower terms}$$

where $g=(g_{ik})$ is a symmetric $n\times n$ matrix with Lorentz signature, one plus and the rest minus, and the indices of u indicate second order derivatives with respect to the variables $(x_1,\ldots,x_n)$. The light rays of the wave equation are replaced by the extremals of the indefinite metric corresponding to g. If $d(x,y)$ denotes the square of the corresponding distance from x to a given point y, $d(x,y)=0$ is the equation of a two-sheeted conoid with its vertex at y. For a given half H of the corresponding cone, Hadamard constructed a fundamental solution $E(x,y)$ with pole at y, $PE(x,y)=\delta(x-y)$, with the following properties. It vanishes outside H and behaves inside H as in the constant coefficient case except that the vanishing inside the conoid when n is even (i.e. odd in the previous notation) is replaced by a smooth behavior up to the boundary. Hadamard guessed the shape of the fundamental solution in the form of an asymptotic series. To verify that the construction yielded a fundamental solution, Green's formula had to be used. This led to difficulties with the singularities on the cone were avoided by a limiting procedure called the method of the finite part. A few years later, Marcel Riesz (1937 and 1949) managed to replace it by a more palatable analytic continuation with respect to a parameter.

It seemed hopeless to extend Hadamard's method to higher order hyperbolic equations. Even an existence proof for Cauchy's problem and with it the existence of fundamental solutions presented problems. Petrovsky managed a very complicated existence proof in the thirties, but an easier one using functional analysis was found in the fifties by Gårding. Still, an analysis of the singularities of the fundamental solutions remained.

The break-through came in 1957 with a paper by Lax. He found out how to make the Fourier method work for variable coefficients by using general oscillatory integrals, ignoring low frequencies and keeping the high ones which are responsible for the singularities. The method was not new in itself. It had been used by physicists under the name of the geometrical optics approximation and, in quantum physics, the semi-classical approximation. But its use for hyperbolic systems and operators of high order was a novelty. The constructions of Hadamard and Lax shared one defect not present in the existence proofs: both were restricted to a neighborhood of the pole of the fundamental solution.

Lax's paper was one of the first that aroused the interest of mathematicians in the analysis of singularities of oscillatory integrals. The outcome has been a vast theory of prime importance whose ingredients are pseudodifferential operators, the notion of wave front set or singularity spectrum for an arbitrary distribution or hyperfunction and a theory of propagation of singularities. Its name is microlocal analysis and it has a host of applications. In one of them, Hörmander and Duistermaat succeded in making the construction of Hadamard and Lax global. The theory of microlocal analysis including many recent results are to be found in Hörmander's books The Analyis of Linear Partial Differential Operators I-IV (Springer 1983-85).

The aim of this series of lectures is to present the use of microlocal theory in the analysis if singularities in linear wave propagation, in the majority of cases represented by the fundamental solutions of linear hyperbolic partial differential and pseudodifferential operators. Chapter 1 deals with forward fundamental solutions of hyperbolic differential operators wit constant coefficients. It presents the theory of lacunas in a general form and has one application to a general form of conical refraction. Chapter 2 about oscillating integrals and wave front sets, Chapter 2 about pseudodifferential operators and Chapter 4 about symplectic geometry present known material necessary for the sequel dealing with the singularities of fundamental solutions of strongly hyperbolic operators and oscillating integrals in general. In Chapter 5 there is a new simple construction of a global parametrix of the fundamental solution of a first order pseudodifferential operator. This construction is basic since parametrices of strongly hyperbolic differential operators are sums of such parametrices paired in a certain way. The final chapters 6 and 7 give a detailed analysis of the singularities of such paired oscillatory integrals.

These lectures were delivered in April and May of 1986 at the Mahematics Institute of Nankai University, Tianjin. The author wants to take this opportunity to thank the Institute for its hospitality and his audience for its patience.

References to the introduction in historical order

Huygens C. Abhandlung über das Licht. Ostwalds Klassiker 20(1913)
Fresnel A.J. Ouevres complètes (1866-70). Imprimerie Impériale Paris.
Lame G. Leçons sur la théorie mathématique de l'elasticité des corps solides. Deuxième edition (1866), Paris, Gauthier-Villars.
Kovalevsky S.V. Über die Brechung des Lichtes in cristallischen Mitteln. Acta Math. 6 (1985)249-304
Tedone O. Sull'integrazione dell'equazione... Ann. di Mat. Ser. 3 vol. 1 (1889)
Volterra V. Sur les vibrations lumineuses dans le milieux biréfringents. Acta Math. 16(1892)154-21
Zeilon N. Sur les équations aux dérivées partielles à quatre dimensions et le problème optique des milieux biréfringents I,II. Acta Reg. Soc. Sc. Upsaliensis Ser. IV vol.5 No 3 (1919),1-57 and No 4(1921),1-130
Hadamard J. Le problème de Cauchy et les équations aux derivées partielles hyperboliques. Hermann et Cie, Paris 1932 .(Originally lectures at Yale Univ. 1922)
Herglotz G. Über die Integration linearer partieller Differentialglechungen mit konstanten Koeffizienten. Ber. Sächs. Akad. Wiss, 78 (1926), 93-126, 287-318, 80 (1928)69-114
Petrovsky I.G. Über das Cauchysche Probelm fur Systeme von partiellen Differentalgleichungen. Mat. Sb. 2(44) (1937)815-870
- On the diffusion of waves and lacunas for hyperbolic equations. Mat. Sb. 17(59) (1945)145-215
Riesz M. L'intégrale de Riemann-Liouville et le problème de Cauchy. Acta Math 81 (1949) 1-223.
Gårding L. Solution directe du problème de Cauchy pour les équations hyperboliques. Coll. Int. CNRS Nancy 1956, 71-90
Lax P. Asymptotic solutions of oscillatory initial value problems. Duke Math. J. 24 (1957) 627-646
Ludwig G. Conical refraction in Crystal Optics and Hydromagnetics. Comm. Pure Appl Math XIV (1961)113-124
Duistermaat J.J. and Hörmander L. Fourier integral operators II. Acta Math. 128 (1972) 183-269
Atiyah M.F., Bott R., Gårding L. Lacunas for hyperbolic differential operators with constant coefficients I,II. Acta Math. 124(1970),109-189 and 131(1973)145-206

CHAPTER 1

HYPERBOLIC OPERATORS WITH CONSTANT COEFFICIENTS

Introduction The main object of this chapter is to express the fundamental solutions of homogeneous hyperbolic differential operators as integrals of rational forms over certain cycles. This yields the Petrovsky condition for lacunas. The first step is a section on algebraic hyperbolicity. In a second section inverses of hyperbolic polynomials are studied. The third section deals with intrinsic hyperbolicity, the fourth with fundamental solutions and in the fifth a formula by Gelfand is used to derive the desired results.

1.1 Algebraic hyperbolicity

Let $f(x)=f(x_1,\dots,x_n)$ be analytic for small x and let $a\neq 0$ be fixed in R^n.

Definition The function f is said to be microhyperbolic with respect to a if

$$\operatorname{Im} t > 0 \Rightarrow f(x+ta) \neq 0$$

for all sufficiently small t and all sufficiently small real x.

Let us develop f at x=0 in series of terms of increasing homogeneity,

$$f(x) = f_0(x)+f_1(x)+\dots+f_m(x)+\dots .$$

The first non-vanishing term, say f_m, is called the principal part of f and will be denoted by Pr f.

Examples. When m=0, $f(0)\neq 0$ and f is trivially microhyperbolic with respect to any a. When m=1 and f is real, f is locally hyperbolic with respect to any a with Pr $f(a)\neq 0$.

Lemma Put $h(t,s)= f(ta+sx)$ with small complex t and s. Then, if f is microhyperbolic with respect to a,

(1.1.1) $$h(t,s) = H(t,s) \prod (t+ d_k(sx,a))$$

where $H(t,s)$ is analytic at the origin, $H(0,0)\neq 0$ and the d_k are analytic for small s and vanish when s=0. If

(1.1.2) $$d_k(sx,a) = c_k(s) + \text{higher terms},$$

the numbers c_k are real and the principal part of $h(t,s)$ is

(1.1.3) $$H(0,0) \prod (t+c_k s) = \Pr f(ta+sx).$$

Note. When f is microhyperbolic with respect both a and -a, it is said to be locally hyperbolic with respect to a. It follows from (3) that Pr f has this property when f is microhyperbolic with respect to a. When f is locally hyperbolic with respect to a, the numbers d_k are real for real x and s and hence $f(x)/\Pr f(a)$ is real.

Proof. Choose an x such that Pr $f(x)\neq 0$. Then the principal parts of f and $h(t,s)$ are the same. Disregarding the definition of m, let m be the least k for which $g_k(0)\neq 0$ in the expansion

(1.1.4) $$h(t,s) = g_0(s) + tg_1(s) +\dots t^k g_k(s)+\dots$$

for small s and t. Without loss of generality, we may also assume that m>0. Then, by the properties of power series, (1) holds, the d_k being Puiseux series in s. But since $h(t,s)\neq 0$ when s is real and t is small with Im t >0, these series are actually power series. In fact, the existence of a first term of the Puiseux series with a fractional exponent is easily seen to contradict this assumption. Hence the degree of Pr $h(s,t)$ is m and equal to that of Pr f. In particular, $H(0,0) = \Pr h(1,0) = \Pr f(a)$ does not vanish, a statement independent of the assumption that Pr $f(x)$ does not vanish. We can then go to (4) again without requiring that Pr $f(x)\neq 0$ and are then sure that $g_m(0)$ does not vanish so that (1) and (2) follow, the formula (3) being a consequence of these two. This finishes the proof.

Let us note that

(1.1.5) $$\mathrm{Pr}\, f(x) = H(0,0)\prod c_k.$$

for all x. In the sequel we shall assume that $\mathrm{Pr}\, f(a)=1$.

Definition Let $C(f,a)$, called the hyperbolicity cone of f, be the component of the complement of the real hypersurface $\mathrm{Pr}\, f(x)=0$ which contains a.

According to (5) this means that x is in $C(f,a)$ precisely when all $c_k = c_k(a,x)>0$ on $C(a,f)$. In fact, when $x=a$, all the numbers c_k are 1.

Theorem $C(f,a)$ is an open convex cone. If K is a compact part of it, f is uniformly locally hyperbolic with respect any b in K. More precisely, there is a positive number A such that

(1.1.6) b in K, $|s|, |x| <A$, $\mathrm{Im}\, s>0 \Rightarrow f(x+sb) \neq 0$.

Proof. Let us write the formula (2) with x replaced by x+sb and with ta and sb interchanged,

$$f(ta+sb+x)= H(t,s,x) \prod (s + d_k(b,x+ta)).$$

Here $H(t,s,x)$ does not vanish for sufficiently small arguments. Further, since the left side does not vanish when s is real and $\mathrm{Im}\, t>0$, none of the numbers d_k crosses the real axis. Hence, since

$$d_k(a,sb) = c_k\, s + \text{smaller}$$

where the c_k are positive, we have

$$\mathrm{Im}\, t>0 \Rightarrow \mathrm{Im}\, d_k(b,x+ta) > 0.$$

when x and s are small enough. Hence the second part of the theorem follows. To prove the first part, note that

$$\mathrm{Pr}\, f(ta+sb) = \mathrm{Pr}\, f(a) \prod (t +sc_k(a,b)).$$

It follows that $C=C(a,f)$ contains $ta+sb$ when b is in C and $t,s >0$. This completes the proof.

Translates

For small real y, let

$$f_y(x) = f(x+y)$$

be the translate of f by y. Our last theorem has the following corollary

Theorem . If f is microhyperbolic with respect to a, so is f_y. The function

$$y \to C(f_y,a)$$

is inner continuous in the sense that if y tends to z, then the right side above contains any compact subset of C(f,a) when z is sufficiently close to y.

Proof. It suffices to prove the theorem when z=0 in which case it follows from the previous one.

Homogeneous hyperbolic polynomials.

When f(x) has a principal part Pr f of order m, then

$$r \to 0 \Rightarrow r^{-m}f(rx) \to \Pr f(x).$$

It follows from this that, if f is microhyperbolic with respect to a, then P(x) = Pr f(x) is a polynomial, homogeneous of order m with the property that

(1.1.7) $\quad \operatorname{Im} s>0,\ x \text{ real} \Rightarrow P(sa+x) \neq 0.$

Such polynomials are said to be hyperbolic with respect to a. The set of those will be denoted by Hyp(a,m). Note that since P is homogeneous, (7) holds with Im s>0 replaced by Im s≠0 so that P is also hyperbolic with respect to -a. It is obvious that if two homogeneous polynomials P,Q are hyperbolic with respect to a, then PQ has the same property and $C(PQ,a)=C(P,a) \cap C(Q,a)$.

Examples. Let m be the degree of a real homogeneous polynomial P. If m=0, P is hyperbolic if and only if P≠0 and then it is hyperbolic with resepect to any a≠0. When m=1, P is hyperbolic with respect to any a with P(a)≠0 and C(P,a) is the half-space P(x)≠0 containing a. The only quadratic forms P which are hyperbolic and not products of linear factors are those such that ± P has a signature with one plus and the rest minus or zero, i.e. in normal form,

$$P(x)= x_1^2-x_2^2-...-x_k^2, \quad 1<k\leq n.$$

Such a P is hyperbolic with respect to any a with P(a)>0 and C(P,a) is the part of the double cone P(x)>0 containing a.

Collecting some of our earlier results, we have

Theorem A homogenous polynomial P, hyperbolic with respect to a, is also hyperbolic with respect to any b in C(P,a). For any y, the function x->P(x+y) is locally hyperbolic with respect to a and, if $C_y(P,a)$ is the corresponding hyperbolicity cone, the function y-> $C_y(P,a)$ is inner continuous.

The cone $C_y=C_y(P,a)$ will be called the local hyperbolicity cone of P at y and the principal part of x-> P(x+y), defined by $P_y(x)$ = lim $s^{-k}P(y+sx)$, s->0, where k is the order of y as a zero of P, will be called the localization of P at y.

Examples. When k=0, $P_y=P(y)$ is a constant and C_y is $R^n\setminus 0$. When m=1, $P_y(x)$ is linear and C_y is a half-space. When k=2, $P_y(x)$ has degree 2 and C_y is a cone or a half-space.

Lineality

The lineality L(P) of a polynomial P is defined to be the set of all y for which P(x+ty)=P(x) for all x and all numbers t. The lineality is a linear space and if its dimension is k, P is a polynomial on the quotient $R^n/L(P)$, a linear space of dimension n-k where k= dim L(P). Very simply one can say that P is a polynomial in n-k but no less number of variables. A polynomial whose lineality vanishes is said to be complete. When P is homogeneous and hyperbolic, $C(P,a)+L(P)= C(P,a)$ trivially. Hence a hyperbolicity cone is open if and only if the corresponding polynomial is complete. When P has a zero of order k at y, $P(x+y)= P_y(x)+$ higher terms in x, P_y not zero as a polynomial in x, we have

$$P_y(x+y)= \lim s^{-k}P(y+s(x+y)) = P_y(x).$$

Hence y is in the lineality of P_y and hence the local hyperbolicity cone at y is invariant under translations in the y direction.

1.2 Distributions associated with the inverses of a homogeneous hyperbolic polynomial

It is well known that if f(z) is analytic in the upper half-plane and $O(y^{-N})$ there for some N, then the limit

$$f(x+i0) = \lim f(x+iy) \text{ for } y \searrow 0$$

exists as a distribution of order < $N+1+\epsilon$ for any $\epsilon>0$. An easy proof depends on a passage to the limit, $b\searrow 0$, in Green's formula in the following form (Hörmander 1983 I, 64),

(1.2.1) $$\int f(x+i(b+c))g(x,c)dx - \int f(x+ib)g(x)dx =$$

$$= -i \int_B f(x+i(y+c))g_{\bar z}(x,y)dxdy$$

Here B is the strip $0<b<y<c$, g(x) is a smooth function with compact support and g(x,y) a smooth extension of g for $0\leq y\leq c$ and $g_{\bar z}$ is defined by the formula $dg=g_z dz+g_{\bar z}d\bar z$. The proof of the formula is easy. Since $f_{\bar z}=0$, we have

$$df(x+i(y+c))g(x,y)dz = 2i\ f(x+i(y+c)g_{\bar z}(x,y)dxdy.$$

Note that the differential vanishes when g is analytic and $g(x,y)=g(x+iy)$. In the general case one defines g(x,y) by the following formula whose right side is g(x+iy) when g is a polynomial of degree at most N,

$$g(x,y) = \Sigma\, g^{(k)}(x)\ (iy)^k/k!,\ k<N+1.$$

A passage to the limit in Green's formula then gives a convergent integral on the right and this proves the statement.

Let us now pass to several variables (Hörmander 1983 I, 66),

Lemma Suppose that $f(z)$, $z=x+iy$ in C^n, is analytic when when z belongs to an open subset of R^n and y to an open cone with its vertex at the origin and belonging to a ball $|y|\lesssim c>0$. Suppose also that

$$|f(x+iy)|\lesssim \text{const } |y|^{-N}.$$

Then the limit

$$f(x+i0) = \lim f(x+iy),\ y\to 0,$$

exists as a distribution.

Proof. Change variables so that the y_1-axis points into the cone, apply the formula (1) to the first variable with the others as parameters and integrate with respect to $x_2,\ldots,x_n$.

Using the lemma we can now introduce two inverses outside the origin of a homogeneous polynomial $P(x)$ of degree m, hyperbolic with respect to a vector $a\neq 0$ in R^n.

Theorem . Let $x\to c(x)$ be a smooth function from $R^n\backslash 0$, homogeneous of degree 1 and chosen so that

$$c(x) \text{ belongs to } C_x(P,a)$$

for all x. Then the limits

$$1/P(x+\epsilon i0) = \lim P(x+isc(x)) \text{ as } 0<s\to 0,\ \epsilon=1 \text{ or } -1,$$

exist independently of the choice of $c(x)$ and define distributions outside the origin which are homogeneous of degree $-m$.

Proof. Since $1/P(x+iy)$ satisfies the hypothesis of the preceding lemma locally, the existence of the two inverses follows by a partition of unity. The homogeneity follows from the homogeneities of P and c(x). Note that, since P is homogeneous, P has unique restrictions to manifolds transversal to radii from the origin in R^n.

1.3 Intrinsic hyperbolicity

When P(D), with $D=\partial/i\partial x$ the imaginary gradient, is a constant coefficient differential operator, let $P(\xi)$ be its characteristic polynomial. A distribution E(x) is said to be a fundamental solution of P if

(1.3.1) $$P(D)E(x) = \delta(x).$$

Definition P is said to be intrinsically hyperbolic if it has a fundamental solution E with support in a proper closed cone K with its vertex at the origin.

Note. By the theorem of supports for convolutions with supports in a cone, such an E is unique.

A linear function t(x) which is positive on $K\backslash 0$ will be called a time function. The intuitive meaning of this definition is that a shock at x=0 in an elastic medium whose movements are governed by P as a physical law should propagate with finite velocity in all space directions, i.e. the directions in the hyperplane t(x)=0. In view of this, K will be called the propagation cone of P.

Without restriction we may assume K to be convex. It is convenient to introduce an open cone C dual to K and defined as the set of time functions. We write them in the form

$$x.\xi = x_1\xi_1+\ldots+x_n\xi_n,$$

required to be invariant under real linear transformations. It is clear that C is open and convex. It will be called the hyperbolicity cone.

Theorem P is intrinsically hyperbolic with cone K if an only if, given a compact subset B of C, there is a continuous real function $s(\eta)=s(-\eta)$ defined in C and homogeneous of degree 1 such that

(1.3.2) $\quad \eta$ in C, $s(\eta)>1 \Rightarrow |P(\xi-i\eta)^{-1}| \lesssim \text{const}(1+|\xi-i\eta|)^{\text{const}}$.

When this condition is satisfied, E is given explicitly as an inverse Fourier-Laplace transform

(1.3.3) $\quad E(x) = (2\pi)^{-n} \int \exp x.\zeta\ P(\zeta)^{-1} d\zeta$

where $\zeta=\xi+i\eta$, η is in $-C$ and $s(\eta)>1$. Under these conditions, the right side is independent of η and the support of E is contained in K.

Note. If (1) is weakened to (4) below and the chain of integration in (3) is modified accordingly, the theorem holds when $P(\zeta)$ is the Fourier-Laplace transform of a distribution $f(x)$ with compact support and (1) is replaced by the convolution $f*g(x)=\delta(x)$.

Proof. Let η be in C and denote the time function $\eta.x$ by $t(x)$. Let $h(x)$ be a smooth function which is 1 when $t(x)<1$ and 0 for $t(x)>2$. We have

$$\delta(x) = P(D)h(x)E(x) + P(D)(1-h(x))E(x)$$

where $g(x)=h(x)E(x)$ and the second function above, say $k(x)$, have compact supports. Let

$$G(\zeta) = \int \exp -ix.\zeta \quad g(x)\, dx$$

be the Fourier-Laplace transform of g and K that of k. We have

$$1 = P(\zeta)G(\zeta)+K(\zeta).$$

The lower bound of $\eta.x$ on the support of k for η in C is a positive, continuous function $u(\eta)$ of η which is homogeneous of degree 1. For η in $-C$ we have

$$|K(\zeta)| \lesssim A(1+|\zeta|)^N \exp -u(\eta)$$

for some positive numbers A and N. It follows that $|K(\zeta)|$ is at most 1/2 and hence (3) holds when

$$|u(\eta)|>\text{const}\ \log(2+|\xi|).$$

Here the logarithm is not necessary which follows from a general algebraic theorem, the Seidenberg-Tarski lemma (Hormander 1983 II 364). With a suitable choice of s, this proves the first part of the theorem. To prove the second part, let E be defined by (3) in the distribution sense and let $f(x)$ be a smooth function with compact support and the Fourier-Laplace transform $F(\zeta)$. Then

$$\int f(-x)E(x)dx = (2\pi)^{-n} \int F(\zeta)P(\zeta)^{-1}d\zeta.$$

Since F is of fast decrease, the integral is absolutely convergent. By Cauchy's theorem, the chain of integration, $\zeta=\xi+i\eta$ with η fixed in $-C$, $s(\eta)>1$, does not infleunce the integral as long as η satisfies the conditions stated. Canging f to $P(-D)f$ proves that E is a fundamental solution. When x is outside K, we can find a permitted η for which $x.\eta>0$. With the support of f close to x, we are then in a situation when the second integrand above tends to zero uniformly when η is replaced by $t\eta$ and t tends to plus infinity. This proves that the support of E is contained in K and finishes this sketchy proof.

Non-homogeneous hyperbolic polynomials

The condition (2) can be simplified considerably. It suffices that, for some time function $\theta.x$, the following condition holds where $P(\xi)$ has degree m and its principal part is denoted by P_m,

(1.3.4) $P_m(\theta)\neq 0$ and $P(\xi+t\theta)\neq 0$ when $\text{Im}\ t <$ some t_0.

When this condition is satisfied, we say that P is hyperbolic with respect to θ. Since $Q=P_m$ is the principal part of P, we have

$$s\to\infty \Rightarrow s^{-m}P(s(\xi+t\theta))\to Q(\xi+t\theta)$$

and it follows that (4) holds for Q with $t_0=0$. Hence Q is also hyperbolic wit respect to θ and the algebraic theory of section 1 applies. In particular, Q is hyperbolic with respect to any η in the hyperbolicity cone $C(Q,\theta)$ of Q. It can be shown that also P has this property but since homogeneous hyperbolic polynomials are our main interest, we shall here leave the non-homogeneous case.

1.4 Fundamental solutions of homogeneous hyperbolic operators. Propagation cones. Localizations. General conical refraction.

When $P(\xi)$ is a homogeneous polynomial which is hyperbolic with respect to θ, let $C=C(P,\theta)$ be its hyperbolicity cone. The dual cone $K=K(P,\theta)$ consisting of all x for which $x.C\geq 0$, i.e. $x.\xi\geq 0$ for all ξ in C, is obviously closed and convex. It will be called the propagation cone of P. The reason is that, according to the preceding section, P has the fundamental solution

$$E(x) = (2\pi)^{-n} \int \exp ix.\zeta \, d\xi/P(\zeta), \quad \zeta=\xi-i\theta,$$

with support in K. In particular, P is intrinsically hyperbolic.

To give an idea of propagation cones, here are some examples where m is the degree of P, supposed to be a real polynomial.

Examples. When $m=0$, P is a constant $\neq 0$, θ may be anything, $C=R^n$ and $K=\{0\}$. When $m=1$, P is hyperbolic with respect to θ if and only if $P(\theta)\neq 0$, C is the half-space $P(\theta)P(\xi)>0$ and K the half-line generated by $P(\theta)\text{grad } P(\theta)$. When $m=2$, P is a quadratic form and it is easy to see that P is hyperbolic with respect to some θ if and only P or $-P$ has Lorentz signature, with one plus and the rest minus or zero. In suitable coordinates we then have

$$P=c^2\xi_1^2-\xi_2^2 - \ldots -\xi_k^2,$$

and $\theta=(1,0,\ldots,0)$. Here C is the open cone $x_1>0$, $P(\xi)>0$ and K is the closed cone $x_1\geq 0$,

$$c^{-2}x_1^2 -x_2^2-\ldots -x_k^2 \geq 0$$

intersected by the hyperplanes $x_{k+1}=0,\ldots,x_n=0$.

Our last example illustrates the fact that the propagation cone K of a hyperbolic polynomial P is orthogonal to the lineality L of P. In fact, the hyperbolicity cone C has the property that $C+L=C$ so that $x.L=-x.L \geq 0$ when x is in K. Since the hyperbolicity cone of the product of two polynomials is the intersection of their hyperbolicity cones, the propagation cone of the product is the union of their propagation cones.

Localization. The wave front surface.

Let Q be the localization of a hyperbolic polynomial P at a point η so that

(1.4.1) $$P(t\eta+\zeta) = t^{m-k}Q(\zeta) + O(t^{k-1}),$$

where k is the order of η as a zero of P. We shall see that the fundamental solution $E(x)$ of P has a corresponding localization. In fact, let t be large real and consider

$$(2\pi i)^n \exp -ix.t\eta\, E(x) = \int \exp i(\zeta-t\eta)d\xi/P(\zeta) =$$

$$=\int \exp i\zeta.x\, d\xi/P(t\eta+\zeta)$$

where $\zeta=\xi-i\theta$. Multiplying by a smooth function $h(x)$ and integrating we get

(1.4.2) $$(2\pi i)^n \int E(x)h(x)\exp -itx.\eta\, dx = \int H(-\zeta)d\xi/P(t\eta+\zeta),$$

where

$$H(\zeta) = \int \exp -ix\zeta\, h(x)dx$$

is the Fourier-Laplace transform of h. Since P is hyperbolic, $|P(t\eta+\zeta)| \geq |P(i\theta| \neq 0$ and, since g is smooth with compact support, $G(-\zeta)$ is fast decreasing in ξ. Hence, using (1), we can pass to the limit $t\to\infty$ and get

(1.4.3) $$t^{m-k} \int \exp -ix.t\eta\, E(x)h(x)dx \to \int F(x)h(x)dx$$

where

$$F(x) = (2\pi)^{-n} \int \exp ix.\zeta\, d\xi/Q(\zeta)$$

is the fundamental solution of Q with support in a propagation cone $K(\eta)$ dual to the hyperbolicity cone $C(\eta)$ of the localization Q of P at η. The propagation cone $K(\eta)$ is said to be local. The union of all $K(\eta)$ for $\eta\neq 0$ is called the wave front surface $W=W(P,\theta)$ of P (with respect to θ). Its properties are given in the following

Lemma . The wave front surface is a closed semialgebraic part of the propagation cone of codimension 1.

Proof. It suffices to consider complete polynomials P. Let $C(\eta)$ be the hyperbolicity cone of the localization of P at η. Since every $C(\eta)$ contains the hyperbolicity cone C of P and the function $\eta \to C(\eta)$ is inner continuous, the local hyperbolicity cones $K(\eta)$ are contained in K and the function $\eta \to K(\eta)$ is outer continuous, i.e. as η tends to ζ, the cones $K(\eta)$ come arbitrarily (conically) close to $K(\zeta)$. This proves that W is a closed part of K and since η is contained in the lineality of $C(\eta)$, $K(\eta)$ lies in a hyperplane when $\eta \neq 0$. This finishes the proof.

Note. If $P(\eta) \neq 0$, $K(\eta)$ is just the origin and if η is a simple zero of P, $K(\eta)$ is a half-line spanned by a multiple of grad P. Hence most of the non-trivial local propagation cones are just half-lines. It can be shown that W is contained in the dual of the hypersurface $P(\xi)=0$, i.e. the hypersurface parametrized by grad $P(\xi)$ with $P(\xi)=0$ which includes the hyperplanes $x.\xi=0$ when ξ is a non-simple point of $P(\xi)=0$.

In Figure 1 , where n=3, some real hypersurfaces P=0 are paired with the corresponding wave front surfaces, both in projective clothing.

Let us now return to the formula (3). If y is a point in K outside W and h is a smooth function with support y but outside W, all the right sides of (3) vanish. Hence the formula shows that the Fourier transform of $E(x)h(x)$ tends to zero for large values of the argument. In the next section, this very weak conclusion will be strengthened to the effect that the Fourier transform has fast decrease and hence that $E(x)$ is smooth (actually real analytic) in K outside W. If we

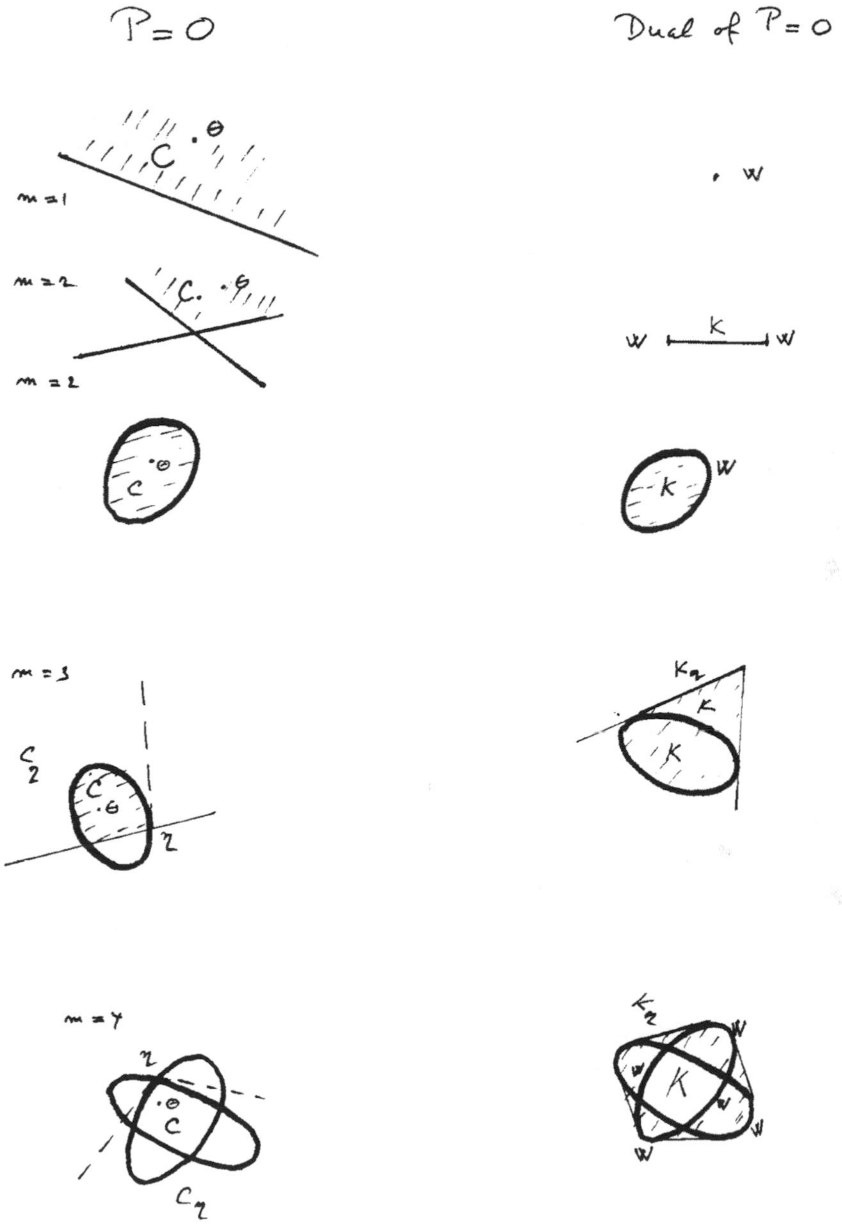

Fig. 1. Hyperbolicity cones and propagation cones when n=3. Some local hyperbolicity cones and the corresponding local propagation cones are also indicated.

interpret (3) in the negative sense, taking the function h with support close to the support of F, we can conclude that the singular support of E contains the union of the supports of the fundamental solutions belonging to the localizations of P at points $\eta \neq 0$. When all these supports are the same as the corresponding local propagation cones, this union is the wave front surface. This happens when the local propagation cones are half-lines and in many other cases. Exceptions may occur when a localization of P has a fundamental solution with a lacuna in its propagation cone.

Conical refraction.

Consider the following formula, proved in the same way as (3),

(1.4.4) $t \to \infty \Rightarrow t^{m-k} \int E(x-y)h(y)\exp -iy.t\eta \, dy \to \int F(x-y)h(y)dy$.

This formula is connected with the phenomenon of double refraction in a general way. When v is a smooth function with compact support, the function

$$u(x) = \int E(x-y)v(y)dy$$

has support in K+supp v and solves the equation $P(D)u(x)=v(x)$. If we let one of the variables x represent time, v(x) can be considered as a source S of vibration and u(x) as the resulting vibration of a medium where propagation of waves is governed by the operator P. To imitate a monochromatic source of high frequency located at a small region in space and time, we put $v(x) = h(x)\exp ix.t\eta$ where η is fixed and t is large and h has small support at the origin. The resulting vibration is then the integral of the left side of (4). The formula says that the limit of a vibration with source $h \exp ix.t\eta$ multiplied by a suitable power of t is a vibration with source h of a medium governed by the localization Q of P to η. This is precisely the situation encountered in conical refraction. A ray of monochromatic light entering a double refracting crystal transversally to a side orthogonal to an optical axis gives rise to a monochromatic source inside the crystal which then propagates according to rules of wave propagation in the crystal. Now since light has a very high frequency, only the high frequency limit becomes clearly visible. In the case of the crystal, the localized operator Q is a wave operator and what is seen in crystals is the projection onto space of its propagation cone $K(\eta)$ in space and time. Since the singularities of the corresponding fundamental solution are situated on the boundary of $K(\eta)$ the boundary is much brighter than the inside. In the actual experiment, light from the propagation cone hits a screen where one can see a luminous ring.

Note. Ludwig (1961) deduced conical refraction from Maxwell's equations. The experiment described above can never be realized precisely since it is impossible to have purely monochromatic light. When the experiment is performed very carefully the luminous ring dissolves into two with a small dark ring in between (see Born and Wolf 1975 for references and discussion). A general theoretical explanation for this phenomenon was given by Uhlmann (1982).

1.5 The Herglotz-Petrovsky formula

In this section we shall deduce formulas for fundamental solutions of homogeneous hyperbolic complete polynomials in the propagation cone but outside the wave front surface as integrals of rational forms over cycles. This could be done starting from their expressions as inverse Fourier-Laplace transforms, but a formula of Gelfand expressing the δ-function in terms of plane waves is a more convenient point of departure. It uses the following family of functions in one variable,

(1.5.1) $$H(s,z) = \int e^{-rz} r^{-s-1} dr = \Gamma(-s) z^{s},$$

where s is real <0 and $\operatorname{Re} z > 0$. The Γ-factor is meromorphic with poles when s is an integer $p=0,1,2,\ldots$. We have

$$\Gamma(-p-t)=\Gamma(1-t)/(-p-t)\ldots(-t)$$

so that

$$H(p+t,z) = -(-1)^p z^{p+t}/(p+t)\dots(1+t)t$$

and hence

(1.5.2) $H(p+t,z) = -(-z)^p/p!t + H(p,z) + Q(t)$

where

(1.5.3) $H(p,z) = -(-z)^p(\log z + \Gamma'(1) - 1 - 1/2 - \dots - 1/p)p!$.

Since $H(s,z)$ is undefined when $s=p$, we are free to use (2) and (3) as a definition. Note that

(1.5.4) $t>0 \Rightarrow H(p,tz) = t^p H(p,z) + (-z)^p \log t/p!$,

so that

(1.5.5) $dH(s,z)/dz = -H(s-1,z)$,

$$dH(p,z)/dz = -H(p-1,z) - (-z)^{p-1}/(p-1)!.$$

We shall use $H(s,z)$ as an analytic function of z in the complex plane cut along the negative real axis. Its boundary values from above and below the real axis are distributions which we shall also use. When t is real, $H(s,it)$ is defined as the distribution $H(s,it+0)$.

For the next lemma we shall need some differential forms,

$$\sigma(\xi) = d\xi_1\dots d\xi_n,\ \sigma_k(\xi) = \sigma(\xi) \text{ with } d\xi_k \text{ taken away},$$

and the Kronecker form

$$\omega(\xi) = \Sigma\ (-1)^{k-1}\sigma_k(\xi),\ k=1,2,\dots,n,$$

with the property that

(1.5.6) $df(\xi)\omega(\xi) = (\xi, \partial f/\partial\xi + n)\sigma(\xi)$.

When f is homogeneous of degree $-n$, the right side vanishes. Also, when f is quasihomogeneous of degree $m-n \geq 0$ in the sense that

$$f(t\xi) = t^{m-n}f(\xi) + p(\xi)$$

where p is a polynomial of degree $m-n$, and $g(\xi)$ is homogeneous of degree $-m$, then

$$df(\xi)g(\xi)\omega(\xi) = p(\xi)\sigma(\xi).$$

Lemma (Gelfand). Let $h(\xi)$ be any smooth real positive function homogeneous of degree 1. Then, in the distribution sense,

(1.5.7) $\delta(x) = (2\pi i)^{-n} \int H(-n,-ix.\xi)\ \omega(\xi)$

with integration over $h(\xi)=1$.

Note. This formula appears in Gelfand-Shilov (1958).

Proof. We shall see that Gelfand's formula results by introducing polar coordinates and integrating out radially in

$$\delta(x) = (2\pi)^{-n} \int \exp ix.\xi \quad \phi(\xi).$$

The right side is the limit as $0<\epsilon$ tends to zero of the integral obtained by adding $-\epsilon h(\xi)$ in the exponential. In the resulting integral we replace ξ by $r\xi$ with ξ restricted to $h(\xi)=1$. The result is the integral

$$(2\pi)^{-n} \int \omega(\xi) \int \exp i(x.\xi +i\epsilon)r\ r^{n-1}dr=$$

$$= (2\pi)^{-n}\int H(-n,-i(x.\xi+i\epsilon)\ \omega(\xi).$$

This proves the lemma.

In our next theorem we shall use the notation

$$H_0(p,z)= -(-z)^p/p!$$

when $p\geq 0$ is an integer and zero otherwise. Then the formula (5) can be written as

(1.5.8) $H(p+t,z) = H(s,z) + H_0(s,z)/t + O(t)$

for all z.

Now let $P(\xi)$ be hyperbolic with respect to θ in $R^n\backslash 0$ and homogeneus of degree m. Let $C(P,\theta)$ be its hyperbolicity cone and $K(P,\theta)$ its propagation cone.

Theorem A When P belongs to $Hyp(\theta,m)$, its fundamental solution with support in the propagation cone $K(P,\theta)$ is

(1.5.9) $E(x) = (2\pi)^{-n} \int \omega(\xi)(H(m-n,ix.\xi)/P(\xi-i0\theta)\ -$

$$-\int \omega(\xi)H_0(m-n,ix.\xi)(\log P/mP)(\xi-i0\theta))$$

with integration over $h(\xi)=1$.

Note. The second term on the right is a polynomial of degree m-n and hence vanishes when $m-n<0$. By the theorem of section 1.2, the right side does not change when θ is replaced by any element of $C(P,\theta)$. The formula can be differentiated under the sign of integration. Hence, the right side is a fundamental solution by Gelfand's formula. To prove the theorem it suffices to show that it is supported in $K(P,\theta)$.

Proof. Note that the integrand is only quasihomogeneous when $m-n\geq 0$. To avoid this case, let s be close to 1 and consider the distribution

(1.5.10) $E_s(x)= \quad (2\pi)^{-n}\int \omega(\xi)H(ms-n,ix\xi)(P(\xi-i0\theta)^{-s}$

with integration over $h(\xi)=1$. Since the integrand is strictly homogeneous of degree 0, it is closed and hence the right side is independent of h. When x is outside $K(P,\theta)$, there is an η in $C(P,\theta)$ such that $x.\eta<0$. Hence

$$(2\pi)^n E_s(x) = \lim \int \omega(\zeta)H(ms-n,ix.\zeta)/P(\zeta)^{-s}$$

where $\zeta=\xi-it\eta$ and $t>0$ tends to 0. But since the integrand is closed, the integral is independent of t. Since it homogeneous of degree 0, we can rewrite it with $\zeta = \xi/t -i\eta$ throughout. Hence, letting t tend to infinity it follows that the right side vanishes.

To arrive at the formula (9) in case $m-n=p\geq 0$, consider

$$(2\pi)^n E_{1+t}(x) = \int \omega(\xi)(H(p,ix.\xi)+H_0(p,ix.\xi)m/t+O(t))/P(\xi-i0\theta)^{1+t}.$$

We shall see that E(x) is obtained by taking the limit as $t\to 0$. In fact, the integrand of

$$\int\omega(\xi)\ H_0(p,ix.\xi)/P(\xi-i0\theta)$$

is homogeneous of degree 0 so that the integral does not change of ξ is replaced by $\xi-it\theta$ with t large. But then, by a previous argument it must vanish. Inserting this result into the previous formula and letting $t\to 0$ proves the desired formula (9). That the right side is a fundamental solution follows from (8) and differentiations under the sign of integration.

We shall now see that $E_m(x)$ and hence also $E(x)$ is analytic in the propagation cone outside the wave front surface $W(P,\theta)$ provided P is a complete polynomial so that K has a non-empty interior. The proof uses Theorem A and the following lemma.

Lemma When x is in the propagation cone but outside the wave front surface, there are continuous functions $\xi \to c(\xi)=c(-\xi)$, homogeneous of degree 1 such that

(i) $c(\xi).x<0$ for all ξ,

(ii) $0<t\leq 1 \Rightarrow tc(\xi)$ in $-C(\xi)$ and $P(\xi+itc(\xi)) \neq 0$.

If we want, (i) can also be replaced by

(iii) $0<t\leq 1 \Rightarrow tc(\xi).x= 0$.

Proof. Follows from the outer continuity of the local hyperbolicity cones and the fact that the real planes $x.\eta=0$ meet every such cone. Note that if the properties (i) and (ii) hold for one x, they hold by continuity for all x in a neighborhood.

Definition Let $\alpha(x)$ be the map

$$\xi \to \xi+ic(\xi),$$

from $h(\xi)=1$.

It is clear that $\alpha(x)$ is a cycle homologous to $h(\xi)=1$. We shall use it to move the integration in the formula (9) of theorem A into C^n.

Theorem B . When x is in K but outside W and with integration over $\alpha(x)$ we have

$$E_m(x) = \int \omega(\zeta)H(ms-n,ix.\zeta)/P(\zeta)^m,$$

$$E(x) = \int \omega(\zeta)H(m-n,ix.\zeta)/P(\zeta) +Q(x),$$

where Q is a polynomial of degree m-n. In particular: $E_m(x)$ and $E(x)$

are analytic in x in $K\setminus W$.

Proof. The integrand of (10) is closed and replacing ξ by $\xi+itc(\xi)$ leads into the analyticity domain of the integrand. Hence the first part of the theorem follows by the last theorem of section 1.2. That $E_\bullet$ is analytic follows since $\alpha(x)$ with the properties (i) and (ii) is locally indepedent of x. The proof of the second part is similar with the complication that the integrand of (9) is no longer closed when $m-n\geq 0$. But then its differential is $\delta(\xi)$ times a polynomial which explains the presence of the polynomial Q in the formula above for E(x). The proof is finished

When $m-n<0$, Theorem B expresses the fundamental solution E(x) as the integral of a rational function over a certain cycle $\alpha(x)$. Using the fact that E(x) vanishes outside $K(P,\theta)$, this situation can be achieved also when $m-n\geq 0$. We shall first work with Theorem A.

When q is an integer, let us put

$$M(q,t)= H(q,it)-(-1)^q H(q,-it).$$

so that

(1.5.11) $\quad q\geq 0 \Rightarrow M(q,t)=-\pi i(it)^q(\text{sgn } t)/q!,$

(1.5.12) $\quad q<0 \Rightarrow M(q,t)= i^q(-q-1)!((t-i0)^q - (t+i0)^q)$

Using Theorem A to compute $E(x)-(-1)^q E(-x)$ we get

Theorem C When $x.\theta>0$,

(1.5.13) $\quad E(x) = (2\pi)^{-n} \int \omega(\xi)M(q,x,\xi)/P(\xi-i0\theta).$

In our next theorem, we shall go out into C^n with the formula (13). At the same time we shall be able to operate in complex projective space C^* of n-1 dimensions and to express the fundamental solution as integrals of the (n-1)-form

$$f(m-n,\zeta) =(2\pi)^{-n} (ix.\zeta)^{m-n} \omega(\zeta)/P(\zeta), \quad \zeta=\xi+i\eta,$$

over certain cycles. The form is defined, holomorphic and closed when η belongs to $-C(\xi)$ and η is sufficiently small when outside the global hyperbolicity cone $C=C(P,\theta)$. Let $\alpha(x)^*$ be the image of $\alpha(x)$ in C^* and orient it so that

(1.5.14) $$\omega(\xi)\ x.\xi > 0.$$

Let X^* and P^* be the complex projective hypersurfaces $x.\zeta=0$ and $P(\zeta)=0$ respectively.

To see what the new cycle is, note that the antipodal map $\xi \to -\xi$ maps $\alpha(x)$ into $(-1)^{n-1}\ \alpha(x)$ where the bar denotes complex conjugation. Hence, in complex space, the boundary $\beta(x)$ of $\alpha(x)^*$, called the Petrovsky cycle, appears as

(1.5.15) twice Re X^* detached from Re P^* when n is odd

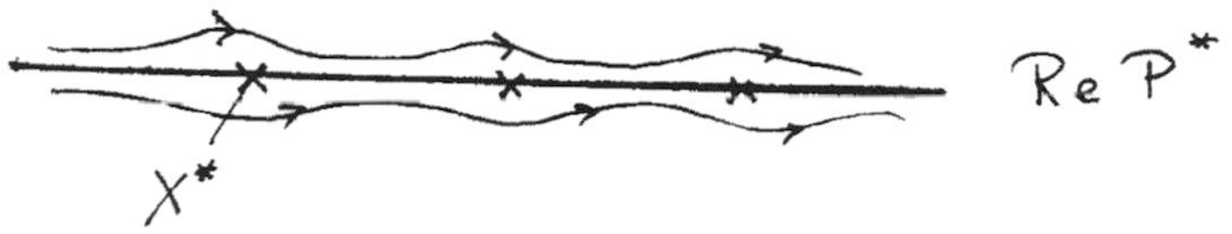

(1.5.15) tubes around Re P^* ψ Re X^* when n is even.

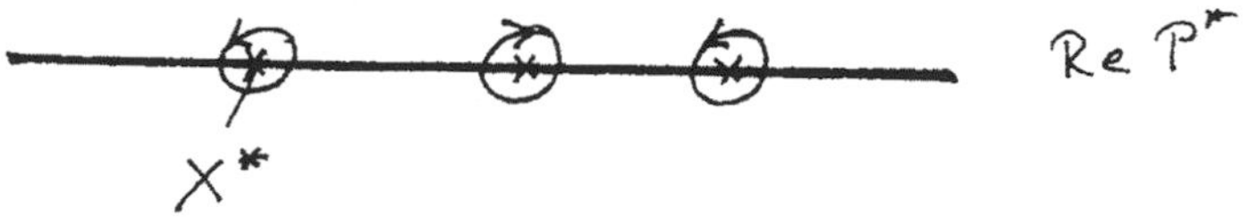

We can now let the integration of Theorem C go out into the complex.

Theorem D. The Herglotz-Petrovsky formula Let P be a complete polynomial hyperbolic with respect to θ and E its fundamental solution with support in the propagation cone $K(P,\theta)$. Then, if $q=m-n$, and x is in K but outside the wave front surface W,

$q\geq 0 \Rightarrow E(x) = -\int \pi i f(q,x.\xi)/q!$, integration over $\alpha(x)^*$.

$q<0 \Rightarrow E(x) \ = \int f(p,x.\xi)$, integration over a tube around $\beta(x)$.

Proof. As the Theorem B, now starting from theorem C.

Since we are dealing with rational forms on complex projective space, and the topology of $P^*\cap X^*$ does not change for x in one component of $K\setminus W$ we have the following result where the Petrovsky cycle appears in a crucial role.

Theorem E. Petrovsky's lacuna theorem . When x is in such a component, E(x) is a polynomial of degree m-n there if the Petrovsky cycle

(1.5.16) $\beta(x)$ is homologous to zero in X^* outside P^*.

Note. Components where of $K\setminus W$ where E(x) is a polynomial (necessarily homogeneous of degree m-n) are called lacunas. The criterion (2) will be called the Petrovsky criterion. It comes very close to being necessary.

The aim of this chapter- to introduce hyperbolicity for constant coefficient operators and to study their fundamental solutions so far as making Petrovsky's lacuna theorem understandable- is now achieved. We round off with an application to the fundamental solution E(x) of the wave operator

$$P(D)= D_1^2-D_2^2-\ldots-D_n^2,$$

hyperbolic with respect to $\theta=(1,0,\ldots,0)$. The propagation cone is defined by

$$x_1\geq 0,\ x_1^2-x_2^2-\ldots-x_n^2 \geq 0,$$

the wave surface is its boundary. When $x=(1,0,..,0)$, Re $X^* \cap P^*$ is empty. Hence, by the Petrovsky criterion, the interior of the propagation cone is a lacuna when n is even and, by its converse, not a lacuna when n is odd.

Note. The Petrovsky criterion has a local variant which can be formulated as follows. Let L be a component of $K\setminus W$ and y a point on its boundary. Let Y be the set of points ξ for which the corresponding local hyperbolicity cones do not meet the real hyperplane $x.\xi=0$. Then $E(x)$ has an analytic extension at y across the boundary of L provided the Petrovsky cycle is homologous in $X^*\setminus P^*$ to a cycle which does not meet Y^*. We shall prove in Chapter 7 that in this form, the criterion survives a passage to hyperbolic operators with smooth coefficients.

Note. The main reference for this chapter is Atiyah-Bott -Gårding (I,1971). The use of Gelfand's decomposition of the δ-function into plane waves (Ge-S 1,p. 118) makes our presentation close to that of Hörmander (1983).

CHAPTER 2

WAVE FRONT SETS AND OSCILLATORY INTEGRALS

The Fourier transform is the supreme tool in the study of partial differential equations with constant coefficients, especially hyperbolic equations. At first sight it seems to be useless for differential equations with variables coefficients. But this is true only for phenomena involving low frequencies. For high frequencies the Fourier transform retains much of its power. One of the first instances when this became apparent was in Lax's construction (Lax 1957) of a parametrix for the fundamental solution of a first order strongly hyperbolic system with variable coefficients. This construction is now part of an established branch of mathematics called microlocal analysis. In this chapter, some of its basic concepts are introduced, the wave front sets and the oscillatory integrals. The chapter ends with Lax's construction applied to a single equation of high order.

2.1 Wave front sets

Let us start with a convenient definition. When C is a closed conical set in R^n with a non-empty interior, a smooth function f defined there is said to be fast decreasing if

$$f(x) = O(|x|^{-N}), \quad |x| \to \infty,$$

for every N>0. Similarly, if C is open and conical, fast decrease means fast decrease in every closed conical subset.

When f is a distribution with compact support, define its regularity set R(f) to be the maximal open conical set where its Fourier transform

$$f^\wedge(\xi) = \int f(x) \exp -ix.\xi \, dx$$

is fast decreasing. When $R(f)$ is $R^n\backslash 0$, f is a smooth function. The complement $S(f)$ of the regularity set will be called the singularity set.

Lemma Multiplication by a smooth function h with compact support does not decrease the regularity set,

(2.1.1) $$R(hf) \supseteq R(f).$$

Proof. The Fourier transform of hf,

$$(hf)^\wedge(\xi) = \int h^\wedge(\xi-\eta) f^\wedge(\eta) d\eta$$

is majorized by

$$c_N \int (1+|\xi-\eta|)^{-N} \; |f^\wedge(\eta)| d\eta$$

for all N and some c_N. Suppose that $f^\wedge$ is fast decreasing in a narrow conical neighborhood C of some ξ_0 and let g be the characteristic function of C and write $f=f_1+f_2$ where f_1 has the Fourier transform $g(f^\wedge)$. Since $g(f^\wedge)$ is fast decreasing, so is $(hf_1)^\wedge$ and it remains to prove that the Fourier transform of hf_2 is fast decreasing when its argument ξ tends to infinity in a some closed conical neighborhood D of ξ_0 contained in C. When verifying this, we may keep η out of C in the integral above with f replaced by f_2. But then $\xi-\eta$ never vanishes and we have

$$|\xi|\geq|\eta| \Rightarrow |\xi-\eta|\geq c|\xi|, \qquad |\xi|\leq|\eta| \Rightarrow |\xi-\eta|\geq c|\eta|$$

for some $c>0$ depending on C and $D\subseteq C$. This leads to the following estimate of $(hf_2)^\wedge$,

$$\text{const} \int (1+|\xi|)^{-N}) \; (1+|\eta|)^M d\eta \; +$$

$$\text{const} \int (1+|\eta|)^{-N}(1+|\eta|)^M d\eta$$

where $|\eta|\leq|\xi|$ holds in the first integral and the opposite inequality in the second one. Here M is fixed and N is arbitrily large. This proves the lemma.

Since the Fourier transform of a derivative

$$D^k f(x), \; D=\partial/i\partial x, \; k=(k_1,\ldots,k_n)$$

is $\xi^k f^\wedge(\xi)$, we have in addition to (1),

(2.1.2) $$R(hD^k f) \supseteq R(f).$$

If u is any distribution and h_1, h_2 are smooth functions with compact supports and supp $h_1 \supset\supset$ supp h_2 are compact, then, by the lemma,

$$R(h_1 u) \supseteq R(h_2 u).$$

Letting the supports of smooth functions h with $h(x) \neq 0$ tend to the point x, the lemma shows that

$$R_x(u) = \lim R(hu)$$

exists independently of the approximating family. The complement $S_x(u)$ of $R_x(u)$ is called the singularity set of u at x. Both are of course sets of rays, i.e. do not contain 0 and with an element ξ they contain all positive multiples of it. The intuitive content of $S_x(u)$ is the set of high frequencies, represented by rays, which are necessary to synthesize u close to x.

Definition (Hörmander) The wave front set WF(u) of a distribution u is the product of its singularity sets,

$$WF(u) = \coprod S_x(\xi).$$

Note. In this way, the singularity set $S_x(u)$ appears as the fiber over x in the wave front set.

For reference we now state.

Lemma Every wave front set is a closed conical subset of $R^n \times R^n \setminus 0$. Differentiation and multiplication by smooth functions do not increase the the fibers of the wave front set of a distribution u, one has

$$S_x(hD^k u) \subset S_x(u)$$

for all x. The projection of the wave front set of a distribution u onto R^n is its singular support.

Examples. Since the Fourier transform of the δ-function $\delta(x)$ is identically one and its support is the origin, the fibers of its wave front set are empty except at x=0 where the fiber is $R^n \setminus 0$. When n=1, we have the decomposition

$$2\pi i\delta(x) = (x-i0)^{-1} + (x+i0)^{-1}.$$

Since the Fourier transforms of the two terms on the right vanish on the negative and positive axes respectively, the fibers over zero of their wave front sets are the positive and negative axis.

It follows from the formula (1.4.3) of the preceding chapter that the wave front set of the fundamental solution $E(x)$ of a homogeneous hyperbolic differential operator $P(D)$ constructed there is contained in the set of points (x,ξ) such that x belongs to $K(\xi)$, the local propagation cone at ξ. In particular, the fiber over 0 is $R^n\backslash 0$, consistent with the fact that $P(D)E(x)=\delta(x)$. In most cases the inclusion is an equality.

Convolutions

The wave front set of the convolution

$$f*g(x) = \int f(x-y)g(y)dy$$

of two distributions one of which has compact support consists of all $(x+y,\xi)$ with (x,ξ) in the wave front set of f and (y,ξ) in that of g. This is easy to prove if we start from the observation that if the Fourier transform of h is the characteristic function of a cone in ξ-space, then, obviously, the wave front set of $f*h$ has all its fibers in that cone.

2.2 The regularity function

A more detailed information on the singularities of a distribution than that obtainable from the wave front set is given by its regularity function. In order to define it, we need some preparation.

Consider the Sobolev spaces H^p, p any real number, of functions with finite Sobolev norm square

$$\|f\|_p^2 = \int |f^\wedge(\xi)|^2(1+|\xi|)^{2p}\, d\xi.$$

We shall see that convolution by a smooth function h with compact support is a continuous self-map of H^p. In fact, if $g(\eta)=(1+|\eta|)^p f\hat{}(\eta)$, then

$$\int(1+|\xi|)^{2p}d\xi\,|\int h(\xi-\eta)f(\eta)d\eta|^2 \leq M\int|g(\eta)|^2 d\eta$$

where M is the maximum of the function

$$\int(1+|\xi|)^{2p}(1+|\eta|)^{-2p}|h(\xi-\eta)|^{2p}\,d\xi.$$

Estimating $|h(\xi-\eta)|$ by $\mathrm{const}(1+|\xi-\eta|)^{-N}$ with large N and using the inequality

$$(1+|\xi|)^{2p} \leq (1+|\xi-\eta|)^{2p}(1+|\eta|)^{2p}$$

when $p>0$ and the same inequality with ξ and η permuted when $p<0$ proves that M is finite. Hence

$$\|hf\|_p \leq C_p\|f\|_p .$$

Next, let C be a conical subset of ξ-space and consider integrals

$$||f||_{p,c} = (\int_c|f(\xi)|^2\ (1+|\xi|)^{2p}d\xi$$

Lemma If B and C are open conical subsets and B is conically relatively compact in C, then

$$\|f\|_{p,c} \text{ finite} \Rightarrow \|hf\|_{p,B} \text{ finite}$$

when h is smooth with compact support.

Proof. Let b and c be the characteristic functions of B and C. By the preceding theorem,

$$\int h\hat{}(\xi-\eta)(1-c(\eta))f\hat{}(\eta)d\eta$$

is fast decreasing in B and, by the preliminaries

$$\|hbf\|_p \leq \mathrm{const}\ \|f\|_{p,B}.$$

The regularity function

The lemmas above show that for every distribution u there are numbers $r_u(x,\xi)$ which express the following property of u: when h is smooth with compact support close to x and $h(x)\neq 0$ and C is a small open cone around ξ, then

$$|\xi|^m(hu)\hat{}\text{belongs to } L^2 \text{ in } C$$

where m comes arbitrarily close to $r_u(x,\xi)$ when the support of h tends to x and the cone C tends to ξ. Of course, $r_u(x,\xi)$ is taken to be the least upper bound of the numbers m. To express this in a natural way, we can say that

$$u \text{ is in } H^r \text{ at } (x,\xi),\ r = r_u(x,\xi).$$

The function $r_u(x,\xi)$ is called the regularity (or singularity) function of u. It is obvious that r_u is homogeneous of degree 0 in ξ, that r_u is identically minus infinity precisely when u is smooth and that the wave front set of u is the closure of the set where r_u is finite.

2.3 Oscillatory integrals.

One of the main objects of microlocal analysis is the study of the singularities of distributions defined by oscillatory integrals, in the simplest case formal integrals

(2.3.1) $$F(x) = \int a(x,\theta) \exp is(x,\theta)\, d\theta$$

with x in an open set X of R^n and θ in R^N which is also the region of integration. The phase function $s(x,\theta)$ and the amplitude function $a(x,\theta)$ are assumed to be infinitely differentiable in $X \times R^N \setminus 0$ and $X \times R^N$ respectively and to have the following crucial properties

(2.3.2) $s(x,\theta)$ is real and homogeneous of degree 1 in θ

(2.3.3) $s_x(x,\theta) \neq 0$ when $\theta \neq 0$,

(2.3.4) $D_x^\alpha D_\theta^\beta a(x,\theta) = O(|\theta|^{m-|\beta|})$ for $\theta \to \infty$.

The last inequality is supposed to hold for a fixed m, called the degree of a, and all α and β, locally uniformly in x. When the terms phase function and amplitude occur in the sequel, they refer to the properties listed above. Note that the product of two phase functions of degrees m and m' is a phase function of degree m+m'.

Since we are only interested in the singularities of F, we shall assume that a vanishes for small θ. Note that (3) means that $s_x(x,\theta)$

is equivalent to $|\theta|$ locally uniformly in x.

We shall see that F is a distribution defined by the formula

(2.3.5) $\int u(x)f(x)dx = \lim \iint a(x,\theta)f(x)\chi(\theta/t) \exp is(x,\theta)\, dx d\theta,\ t\to\infty$,

where f and χ are infinitely differentiable with compact supports and $\chi=1$ in a neighborhood of the origin.

In fact, by the conditions (2) and (3), the differential operator L_x defined by

$$L_x = |s_x(x,\theta)|^{-2}\ \bar{s}_x.\partial/i\partial x$$

exists and reproduces the exponential exp $is(x,\theta)$ and hence the integral of (5) does not change if we apply any power of the adjoint M_x of L_x to the rest of the integrand. Since the coefficients of L_x and M_x are $O(|\theta|^{-1})$, this introduces a new amplitude of arbitrarily large negative degree. Hence the limit of (5) exists regardless of the choice of χ and defines a distribution F in X. Note that, so far, (3) has been used only for $\beta=0$.

If we differentiate the oscillatory integral (1) formally, we get an oscillatory integral with a another amplitude function and it is immediately verified that this operation commutes with the passage to the limit of (5). Hence derivatives of oscillatory integrals exist and are obtained by the usual formal rules. The same holds for multiplication by infinitely differentiable functions.

Conical support, singular support.

It is clear that there are amplitude functions $c(x,\theta)$ which are 1 in a product of a compact set in X and a closed conical set C in R^N and vanish in an arbitrarily small conical neighborhood of $Y\times C$. Every such amplitude function gives rise to a partition of the amplitude a,

$$a(x,\theta)= (1-c(x,\theta))a(x,\theta) + c(x,\theta)a(x,\theta),$$

Let F=G+H be the corresponding oscillatory integrals. If the second amplitude is fast decreasing for all x, G is infinitely differentiable. But this happens also when the gradient $s_\theta(x,\theta)$ does not vanish on the support S of ca. In fact, then its norm is bounded from below and the differential operator

$$L_\theta = |s_\theta(x,\theta)|^{-2}\bar{s}_\theta . \partial_\theta / i$$

is defined in an open conical neighborhood of S. It reproduces the exponential of G and its coefficients and those of its adjoint are infinitely differentiable for $\theta \neq 0$ and homogeneous of degree ≤ 0 in a conical neighborhood of $Y\times C$. Hence it follows from (5) that applying large powers of the adjoint of L_θ to the amplitude ca diminishes its degree by any integer. We conclude that the oscillatory integral G equals another one with an arbitrarily decreasing amplitude and hence is infinitely differentiable.

To formulate the results above in general terms, define the conical support of the amplitude function $a(x,\theta)$ as the complement of the union of sets $Y\times C$ where Y is open in X and C is open and conical in R^N and $a(x,\theta)$ is of fast decrease uniformly for x in compact parts of Y and θ in closed conical parts of C. Let S be the set, closed and conical, where

$$s_\theta(x,\theta)=0.$$

It follows from the above that if $c(x,\theta)$ is an amplitude function which equals 1 for large θ in a neigborhood of the intersection of S with the conical support of $a(x,\theta)$ and vanishes outside another neighborhood, then the oscillatory integral F differs from the oscillatory integral with amplitude $c(x,\theta)a(x,\theta)$ by an infinitely differentiable function. In particular, $s_\theta(x,\theta)=0$ for some non-zero θ when x is in the singular support of F, regardless of the nature of the amplitude function. We can write this observation as

$$\text{sing supp } F \subset \{x, s_\theta(x,\theta)=0,\ x,\theta \text{ in con supp } s(x,\theta)\}$$

The wave front set of an oscillatory integral.

It is now easy to majorize the wave front set of F. Replacing f(x) in (5) by $f(x)\exp{-ix.\xi}$, we get an oscillating integral with amplitude

$f(x)a(x,\theta)$ and phase function $s(x,\theta)-x.\xi$. Keeping θ and ξ away from zero consider its gradient

$$s_x(x,\theta)-\xi \ , \ |\theta|>0.$$

If the rays generated by the two terms (under muliplication by positive numbers) are separate for x in some compact set and θ and ξ in closed conical sets, it is clear that the norm of the gradient is uniformly equivalent to $|\theta|+|\xi|$. Hence, applying large powers of the adjoint of the corresponding L_x to the amplitude function we get a new amplitude function which decreases arbitrarily in θ and ξ in the product of the two conical sets. It follows that

$$\int F(x)f(x) \exp ix.\xi \, dx$$

is fast decreasing for ξ in some conical closed set C provided $s_x(x,\theta)-\xi$ does not vanish for x in a neighborhood of supp f. Combining this with what we know about the singular support of F proves

Theorem. The wave front set of the distribution (1) is contained in the set of pairs (x,ξ) for which

$$\xi = s_x(x,\theta) \text{ and } s_\theta(x,\theta)=0.$$

Note. The first condition says precisely that the rays generated by $s_x(x,0)$ and ξ are the same.

Note. The theorem holds for every amplitude function $a(x,\theta)$. When an individual one is taken into account, we can restrict the wave front set of F further by restricting (x,θ) to the conical support of $a(x,\theta)$.

Examples. The wave front set of the oscillatory integrals

$$\int \exp ix.\xi \, d\xi \quad \text{and} \int \exp i(x-y).\zeta \, d\zeta$$

with ξ and ζ in R^n are, respectively, $(0,R^n \setminus 0)$ and $(x=y, \xi=\zeta, \eta=-\zeta, \zeta$ in $R^n \setminus 0)$.

The wave front set of the oscillatory integral

$$\int \exp i(x.\xi-|\xi|)\, d\xi, \quad \dim x=n,$$

is contained in the set of points (x,ξ) for which $x=\xi/|\xi|$. In particular, its singular support is the unit ball.

It will turn out later that very general examples of oscillatory integrals are those of the form

$$u(x) = \int a(x,\xi) \exp i(x.\xi-H(\xi))\, d\xi$$

where a vanishes when ξ is outside som open conical set C in R^n. The wave front set of u is contained in the set of pairs (x,ξ) with ξ in C for which

$$x=H'(\xi).$$

In geometric language, this means that the hyperplane $x.\eta=1$ touches the hypersurface $H(\eta)=1$ at the point x. In other words, the projection of the wave front set on x-space is a hypersurface dual to the hypersurfaces $H(\eta)=$ const.

It is well known that the duals of very regular hypersurfaces can have very complicated singularities. Suppose for instance that ξ_1 does not vanish on $C\backslash 0$ and put

$$H(\xi) = \xi_1 F(\xi_2/\xi_1,\ldots)$$

where $F=F(t)=F(t_2,\ldots,t_n)$ is any smooth function. That $x=H'(\xi)$ then means that

$$x_1=F(t)-t_2F_2(t)-\ldots, \qquad x_2 = F_2(t),\ldots$$

where F_j equals $\partial F(t)/\partial x_j$. To see what this means, take n=2, $t_2=t$ and

$$F(t) = 1+at+bt^2+ct^3+\ldots .$$

We get

$$x_1=1-bt^2-2ct^3+\ldots,$$

$$x_2 = a+2bt+3ct^2+\ldots .$$

When $b\neq 0$, the dual is approximately a parabola for small t. When b=0, $c\neq 0$, we get a cusp at the origin in the x_1,x_2-plane. This illustrates the well-known result is that the dual hypersurface is smooth unless the Hessian $H''(\xi)$ is degenerate, i.e. has more eigenvectors with the eigenvalue zero than the obvious one, namely ξ (note that $H''(\xi).\xi=0$ since $H'(\xi)$ is homogeneous of degree 0).

2.4 Fourier integral operators.

Associated with oscillatory integrals are the Fourier integral operators

(2.4.1) $$Fu(x) = \int a(x,\theta)u^\wedge(\theta) \exp is(x,\theta)\, d\theta$$

introduced by Hörmander. Here $u^\wedge(\theta)$ is the Fourier transform of a distribution $u(t)$ in R^N with compact support, $s(x,\theta)$ is a phase function and $a(x,\theta)$ an amplitude. The product $a(x,\theta)u^\wedge(\theta)$ fails to be an amplitude only in that (2.3.4) holds only for $\beta=0$. Hence, operating with the differential operator corresponding to s_x proves that $Fu(x)$ is a distribution for every u. When $v(t)$ and $f(x)$ are infinitely differentiable functions with compact supports, the definition (2.3.5) shows that we can write

(2.4.2) $$\int F(uv)(x)f(x) \exp -ix.\xi\, dx$$

as

(2.4.3) $$\int\int f(x)a(x,\theta)(uv)^\wedge(\theta) \exp i(s(x,\theta)-x.\xi)\, dx d\theta,$$

and, if u is infinitely differentiable in a neighborhood of the support of v, as

(2.4.4) $$\int f(x)\exp -ix.\xi\, dx \int\int a(x,\theta)(uv)^\wedge(t)\exp i(s(x,\theta)-\theta.t)\, dt d\theta$$

The first formula shows that (2.4.3) is fast decreasing for ξ in some closed conical set when the rays generated by $s_x(x,\theta)$ and ξ are separate for x in the closure of the support of f. In all this we may of course disregard all θ in open conical sets where $a(x,\theta)(uv)^\wedge(\theta)$ is fast decreasing. The interior oscillatory integral of (2.4.4) acquires, in the usual way, an amplitude of any negative degree provided the gradient $s_\theta(x,\theta)-t$ is different from zero for all θ and for all t,x in the product of the closures of the supports of v and f.

Hence, in that case, it is infinitely differentiable so that (4) is fast decreasing in ξ. Combining all this proves an important theorem about the wave front sets of Fourier integral operators.

Theorem The wave front set of a Fourier integral operator

$$\int a(x,\theta)u^\wedge(\theta) \exp is(x,\theta)d\theta$$

consists of pairs (x,ξ) for which

$$\xi=s_x(x,\theta) \text{ and } (s_\theta(s,\theta),\theta) \text{ is in } WF(u).$$

2.5 Applications

The wave front sets of distributions on manifolds.

To every smooth bijection $x\to f(x)$ of R^n there is a corresponding bijection $u\to v$ of distributions,

$$v(x) = u(f(x))$$

of distributions. When u has compact support, we can express v(x) as a Fourier integral operator

$$v(x) = (2\pi)^{-n}\int u^\wedge(\eta)\exp if(x).\eta \, d\eta,$$

where $u^\wedge$ is the Fourier transform of u. The theorem on the wave front sets of Fourier integral operators has the following application.

Theorem A pair (x,ξ) belongs to the wave front set of a distribution $v(x)=u(f(x))$ if and only if the pair

$$(f(x),{}^tf'(x)^{-1}\xi)$$

belongs to the wave front set of u.

Note. The moral of this result is that the wave front set of a distribution on a manifold X is part of the cotangent bundle $T^*(X)$ of X. In fact, if (y,η) is the pair above, we have $\xi.dx=\eta.dy$.

Proof. By the preceding theorem, the wave front set of $v(x)$ consists of pairs (x,ξ) with $\xi= {}^tf'(x)\eta$ and $(f(x),\eta)$ in the wave front set of

u. Changing the roles of u,v and changing f to its inverse proves that the inclusion is a bijection. This proves the theorem when u ha compact support and hence in general.

Parametrices of fundamental solutions

The first time that Fourier integral operators appeared explicitly was in Peter Lax's construction of parametrices of fundamental solutions of strongly hyperbolic first order systems. The construction extends to strongly hyperbolic differential operators with smooth coefficients

$$P(x,D)=\Sigma\, a_J(x)D^J, \quad |J|\leq m.$$

Here $x=(x_0,\dots,x_n)$ stands for n+1 real variables, $D=\partial/i\partial x$ is the imaginary gradient. The multiindex $J=(J_0,\dots,J_n)$ has n+1 integral components ≥ 0 and D^J, defined accordingly is a derivative of order

$$|J| = J_0+\dots+J_n.$$

That P is strongly hyperbolic means that the characteristic polynomial of its principal part,

$$p(x,\xi) = \Sigma\, a_J\xi^J, \quad |J|=m,$$

is a product of a non-zero function $p_0(x)$ and m factors

(2.5.1) $$p_j(x,\xi)= \xi_0 - q_k(x,\xi)$$

where the q_k are independent of ξ_0 and real and separate for ξ real and not zero. In the sequel we shall put $p_0=1$. As explained for instance in Hörmander 1985 (IV, 394-395), P has a unique fundamental solution E(x) with pole in x=0 which vanishes for $x_0 < 0$. The Cauchy problem with data on the hyperplane $x_0=0$,

(2.5.2) $$PF(x)=0,$$

$$D_0{}^kF=0,\ D_0{}^{m-1}F=i\delta(x_1)\dots\delta(x_n) \text{ when } x_0=0,\ k<m-1,$$

also has a unique solution. By direct computation one finds that

$$E(x)=H(x_0)F(x)$$

where H is the Heaviside function. When

$$P=D_0{}^2-D_1{}^2-\dots -D_n{}^2$$

is the wave operator the function F can be computed explicitly. In fact,

$$F(x) = (2\pi)^{-n} \int ((\exp ix_0 t + ix. \quad - \exp -ix_0 t + ix. \ )/2t \, d\xi$$

where $\xi_0=0$ and

$$t=(\xi_1^2+...+\xi_n^2)^{1/2}.$$

Lax's idea was to imitate this formula by putting, in the general case,

(2.5.3) $F(x) = \Sigma \int a_k(x,\xi)\exp is_k(x,\xi)\, d\xi, \ \xi_0=0,$

with amplitudes a_k and phases s_k determined so that (1) holds in the sense of oscillatory integrals. The success of this approach is formulated in

Theorem If the phase functions s_k are chosen so that

(2.5.4) $p_k(x,s_{kx})=0,$

$$s_k(x,\xi)=x_1\xi_1+...+x_n\xi_n \text{ when } x_0=0,$$

there are formal sums

$$a_k(x,\xi) = \Sigma a_{kj}(x,\xi) \ , \ j=1-m,-m,-1-m,... \ ,$$

whose terms are smooth functions when $\xi=(\xi_1,...,\xi_n) \neq 0$ and homogeneous of degree j in ξ such that the formal sums

(2.5.5) $(\exp -is_k(x,\xi))P(x,D)a_k(x,\xi)\exp is_k(x,\xi))$

vanish of infinite order and the formal sum

(2.5.6) $(\exp -is_k(x,\xi))D_0^k a_k(x,\xi) \exp is_k(x,\xi)$

vanishes when k<m-1 and equals $i(2\pi)^{-n}$ when k=m-1.

Note. Since the a_k are formal sums of strictly homogeneous terms, they are not amplitudes in the technichal sense. However, it will be proved in the next chapter that if $\chi(\xi)$ is a smooth function which is 0 in one neighborhood of the origin and 1 outside another neighborhood and the numbers t_j increase sufficiently fast, then the sums

$$\Sigma\chi(\xi/t_j)a_{kj}(x,\xi)$$

converge to proper amplitudes $b_k(x,\xi)$. If we replace the a_k in (2) by these b_k, we get a distribution G(x) which solves the problem (1) modulo smooth functions. Restricting it to $x_0>0$, we get a true parametrix of the fundamental solution $E(x)$.

Note. Since the p_k are homogeneous of degree 1 in η, the s_k have the same property. The Hamilton-Jacobi differential equations (4) are solvable only for x close to the origin. This means that our parametrix is only local. That a global parametrix exists will be proved in Chapter 5.

Proof. The proof of the theorem is a straightforward verification based on the following

Lemma When $s(x,\eta)$ is a phase function, there are differential operators

$$Q_j(x,\eta,D) \ , \ k=0,\ldots,m,$$

homogeneous of degree j in η such that

(2.5.7) $\quad \exp -is(x,\eta)\ P(x,D)a(x)\exp is(x,\eta) \ = \ \Sigma Q_j(x,\eta,D)a(x)$

for every smooth function $a(x)$. In particular,

$$Q_m = p(x,s_x),$$

$$Q_{m-1} = p^{(j)}(x,s_x)D_j \ +P_{m-1}(x,s_x)$$

where

$$p^{(j)}(x,\eta) = \partial p(x,\eta)/\partial\eta_j$$

and $P_{m-1}(x,\eta)$ is the sum of terms of degree m-1 of the characteristic polynomial $P(x,\eta)$ of $P(x,D)$.

This lemma is proved by a direct verification which is left to the reader. Its formulas inserted into (5) show that the vanishing of all the terms of the formal sum, ordered according to their degrees in η, amounts to linear differential equations called transport equations,

$$L_k(x,\eta,D)a_{kj}(x,\eta) \ = \ b_{kj}(x,\eta)$$

where L_k is the differential operator Q_{m-1} with $s=s_k$ and the right side depends only on the a_{kl} with $l<j$ already constructed. This leaves every a_{kj} free when $x_0=0$. These initial values on the other hand can

be chosen so that the sums (5) get the desired properties. For our principal terms $c_k = a_{kj}$ with $j=1-m$ we get the following equations

$$q_k^j c_k = 0, (j<m-1), \qquad q_k^{m-1} c_k = i(2\pi)^{-n},$$

which, since the q_k are separate, have unique solutions c_k which are homogeneous of degree $1-m$ in η. The other terms a_{kj} are put equal to zero when $x_0=0$. These are the essentials of the proof. The details are left to the reader.

The pairing

The fact that we are dealing with polynomials $P(x,\eta)$ creates a number of special circumstances. The polynomial $p(x,\eta)$ is homogeneous of degree m and also the product of the factors (1). Hence, changing η to $-\eta$ we get pairing $k \to k'$ such that $p_{k'}(x,-\eta) = -p_k(x,\eta)$, in other words

(2.5.8) $$q_{k'}(x,-\eta) = -q_k(x,\eta).$$

If we order the q_k so that $q_1 > \ldots > q_m$, the pairing is simply reflection in the midpoint of the sequence $1,\ldots,m$. Hence there is no and one q_k which is paired to itself if and only if m is even or odd. According to (4), the pairing implies that

(2.5.9) $$s_{k'}(x,\eta) = -s_k(x,-\eta).$$

Since the equations (5) and (6) are invariant under the simultaneous change $k \to k'$ and $\eta \to -\eta$, it is not difficult to see that

(2.5.10) $$a_{k'j}(x,\eta) = (-1)^j a_{kj}(x,-\eta)$$

where $a_k = \Sigma a_{kj}$ is the expansion of a_k in terms of homogeneity j.

The equations (9) and (10) have consequences for the wave front set of F, by our general rules contained in the set of points (x,ξ) for which

$$\xi = s_{kx}(x,\eta) \;, \; \partial s_k(x,\eta)/\partial\eta = 0$$

for some k. By the pairing, both s_k and $s_{k'}$ contribute to the fiber of the wave front set over a given x . Hence the projection on the wave front set of F on x-space appears as [m/2] connected sheets. In particular, for the wave equation there is one sheet and for equations of order 3 and 4 there are two of them. The equations (8) have consequences for the nature of the behavior of the fundamental solution at its singularities. They are responsible, for instance, for the fact that the support of the homogeneous wave equation with constant coefficients in four variables is the light cone and hence coincides with its singular support, a phenomenon called Huygens' principle. In Chapters 6 and 7, the effects of a general pairing for pseudodifferential operators will be investigated.

Note. The material of this chapter appeared for the first time in Hörmander 1971 and has since become standard microlocal analysis. Lax's paper is Lax 1957.

Chapter 3

PSEUDODIFFERENTIAL OPERATORS

Introduction: differential operators and their symbols .

The calculus of pseudodiifferential operators is an extension of that of differential operators with smooth, i.e. infinitely differentiable coefficients

$$a(x,D) = \Sigma a_\alpha(x)D^\alpha , \quad D=\partial/i\partial x ,$$

in some open subset X of R^n. Here $\alpha=(\alpha_1,\dots,\alpha_n)$ are multiindices and $D=(D_1,\dots,D_n)$. Later we shall also use the notation $\alpha! = \alpha_1!\dots\alpha_n!$.

To handle adjoints and products of such operators it is convenient to introduce their characteristic polynomials or symbols

$$a(x,\xi) = \Sigma a_\alpha(x)\xi^\alpha = \exp -ix.\xi \; a(x,D) \exp ix.\xi,$$

with ξ in R^n and $x.\xi = x_1\xi_1+\dots+x_n\xi_n$. It is clear that the map $a(x,D)\to a(x,\xi)$ is a linear bijection. A few moments of reflection show that the adjoint

$$\Sigma D^\alpha \bar{a}_\alpha(x)$$

of a with respect to the sesquilinear form $\int u(x)\bar{v}(x)dx$ has the characteristic polynomial

$$\Sigma\Sigma \; D_x^\beta a_\alpha(x)(iD_\xi)^\beta(\alpha+\beta)!/\alpha!\beta!.$$

Using the multinomial theorem, we can also write it as

$$(\exp iD_x.D_\xi)a(x,\xi).$$

Since $a(x,D)\exp ix.\xi = \exp ix.\xi \; a(x,D+\xi)$, the characteristic polynomial of a product $a(x,D)b(x,D)$ of two differential operators is

$$\Sigma \; a^{(\alpha)}(x,\xi)D_x^\alpha b(x,\xi)/\alpha!$$

where

$$a^{(\alpha)}(x,\xi) = (iD_\xi)^\alpha a(x,\xi).$$

Another way of writing the characteristic polynomial of a product is

$$\exp iD_\eta.D_x \; a(y,\eta)b(x,\xi) \quad \text{for } y=x, \; \eta=\xi.$$

Finally, if $y=f(x)$ is a smooth bijection of X to another open subset Y of R^n, the differential operator $a(x,D)$ transported to Y has the characteristic polynomial

$$b(y,\eta) = \exp -if(x).\eta \; a(x,D) \exp if(x).\eta.$$

We shall see in the next section that all these formulas and also the formulas for changes of variables survive essentially for pseudodifferential operators

$$a(x,D)u(x) = (2\pi)^{-n} \int a(x,\xi)u\hat{}(\xi) \exp ix.\xi \, d\xi,$$

where u is a distribution in X with compact support, $u\hat{}$ is its Fourier transform, $a(x,\xi)$ is an amplitude called the symbol of the operator $a(x,D)$ and the integral is oscillatory. When $a(x,D)$ is a differential operator, the right side is actually $a(x,D)u(x)$ by the Fourier inversion theorem so that the notation is consistent.

Since the ξ gradient of $x.\xi$ is x, the rules for computing the wave front sets of oscillatory integrals shows that pseudodifferential operators do not increase wave front sets, the wave front set of $a(x,D)u$ is contained in that of u.

If we make the Fourier transform of u explicit in the definition of the peudodifferential operator $a(x,D)$, we can write

$$a(x,D)u(x) = \int A(x,y)u(y)dy$$

where the kernel $A(x,y)$ is a distribution defined by the oscillatory integral

$$A(x,y) = (2\pi)^{-n} \int a(x,\xi) \exp i(x-y).\xi \, d\xi,$$

which is smooth in every open set where x is not equal to y.

When the amplitude has degree $-\infty$, i.e. $a(x,\xi) = O(|\xi|^{-N})$ for large ξ and every N, locally uniformly in x, the kernel is a smooth function. It is important to keep in mind that the calculus of pseudodifferential operators to be presented below is a calculus modulo pseudodifferential operators with smooth kernels. In this sense it is a calculus of singularities.

3.1 The calculus of pseudodifferential operators

In order to extend the formulas above from differential operators to pseudodifferential operators, we need some technical information.

We let S^m be the space of amplitudes of degree at most m, i.e.

$$D(\alpha,\beta)a(x,\xi) = O(|\xi|^{m-|\beta|}),$$

locally uniformly for x in compact subsets of X. This notation will also be used for amplitudes where x is replaced by two variables x and y. Topologized by the corresponding seminorms, S^m becomes a Frechet space. First, we shall need a result on asymptotic series $a_1+a_2+\ldots$ of amplitudes a_j whose degrees m_j decrease strictly to minus infinity. When a is an amplitude, we say that

$$a \sim a_1+a_2+\ldots$$

when a is an amplitude and

(3.1.1) $\qquad$ degree $(a-a_1-\ldots-a_j) = m_{j+1}$

for all j, locally uniformly in x. In view of the following lemma, we shall sometimes identify an amplitude and its asymptotic expansion.

Lemma Given $a_1,\ldots$ as above, there is an amplitude a of degree m_1 such that (1) holds.

Proof. By a partition of unity on X, we can restrict x to a compact subset of X. Let $x(\xi)$ be smooth and 0 for $|\xi|<1$ and 1 for $|\xi|>2$. Put

$$b_j = x(\xi/t_j)a_j(x,\xi)$$

and choose the numbers t_j tending to infinity so fast that

$$D(\alpha,\beta)b_j(x,\xi) = O(2^{-j}|\xi|^{m-|\beta|}),\ m=m_{j-1},$$

for $|\alpha|+|\beta|<j+1$. This is clearly possible and can be done so that the sum of the b_j is locally finite and hence defines a smooth function a. Moreover, for all N,

$$D(\alpha,\beta)(a-b_1-\ldots-b_{N+1}) = O(|\xi|^{m-|\beta|}),\ m=m_N$$

when $|\alpha|+|\beta|<N+2$. Since all b_k-a_k are in $\cap S^{-k}$ for $k=1,2,\ldots$ and b_{j+1}

and the following terms are of degree m_{j+1}, (1) follows.

Next, we shall consider pseudodifferential operators defined by amplitudes $b(x,y,\xi)$ depending on x and y,

$$Bu(x) = (2\pi)^{-n}\int b(x,y,\xi)u(y) \exp i(x-y).\xi \, dyd\xi.$$

Lemma If $a(x,\xi)$ is an amplitude with the asymptotic expansion

$$\exp iD_yD_\xi \, b(x,y,\xi)|x=y$$

then $a(x,D)-B$ has a smooth kernel.

Proof. By Taylor's formula we have

$$b(x,y,\xi)= \Sigma((iD_y)^\alpha b(x,x,\xi))(y-x)^\alpha/\alpha! +$$

$$+ \Sigma \ (iD_y)^\alpha c_\alpha(x,y,\xi)(x-y)^\alpha/\alpha!$$

with summations over $|\alpha|<N$ and $|\alpha|=N$ respectively. Here the $c_\alpha(x,y,\xi)$ are amplitudes of the same degree as b. Inserting this into the formula for Bu and performing integrations by parts with respect to ξ, which turns $i(x-y)$ into D_ξ, gives an amplitude

$$\Sigma((iD_y)^\alpha D_\xi{}^\alpha b(x,y,\xi))u(y)dyd\xi/\alpha! \ +\Sigma \ ((iD)_y{}^\alpha D_\xi{}^\alpha c_\alpha(x,y,\xi)u(y).$$

The amplitudes of the last term have the degrees m-N which means that the corresponding kernels are continuously differentiable N-m-n+1 times. The kernel of the first term, on the other hand, differs from the kernel of $a(x,\xi)$ by a kernel of the same smoothness, Hence the kernel of $a(x,D)-B$ is smooth and this proves the lemma.

Polyhomogeneous operators.

A pseudodifferential operator $a(x,D)$ and its amplitude $a(x,\xi)$ of degree m are said to be polyhomogeneous if, as for differential operators, the amplitude is the asymptotic sum for j=0,1,... of amplitudes $a_{m-j}(x,\xi)$ which are homogeneous of degree m-j for large values of ξ. Such amplitudes are of course unique modulo $^{-k}$ for k=1,2,.... . A polyhomogeneous operator $a(x,D)$ with symbol $\sim \Sigma \ a_k(x,\xi)$

has a kind of dual a'(x,D) with symbol

$$a'(x,\xi) \sim \Sigma\ (-1)^k a_k(x,-\xi).$$

When $a(x,\xi)$ is a polynomial, we have $a'(x,D)=a(x,D)$. Such operators are said to satisfy the 'transmission condition' (Hörmander 1985 III p. 110) which guarantees that boundary problems make sense for them. Here we shall use another property, namely that for polyhomogeneous operators the map a->a' is remarkably stable under operations such as taking the adjoint or changing variables.

Adjoints

Theorem The adjoint of a pseudodifferential operators with respect to a sesquilinear duality $(u,v) = \int u(x)\bar{v}(x)dx$, is a pseudodifferentiable operator with the symbol

$$\exp iD_x.D_\xi \bar{a}(x,\xi).$$

Proof. Writing $(u,a(x,D)v)$ explicitly we get

$$(2\pi)^{-n} \int u(y)\bar{a}(y,\xi)\bar{v}(x) \exp i(x-y).\xi\ d\xi dxdy$$

when u an v are smooth with compact supports. Hence the adjoint of a(x,D) is a pseudodifferential operator with the kernel

$$(2\pi)^{-n} \int \bar{a}(y,\xi) \exp i(x-y).\xi\ d\xi$$

so that the desired result follows form the preceding lemma. The explicit form of the symbol of the adjoint shows at once that a->a' commutes with taking the adjoint.

Products

When computing the product of two psududifferential operators with symbols $a(x,\xi)$ and $b(x,\xi)$, we have to suppose that $b(x,\xi)$ vanishes for x outside some compact subset of X.

Theorem If b(x,D) conserves compact supports, the product of two pseudodifferential operators with symbols $a(x,\xi)$ and $b(x,\xi)$ is a pseudodifferential operator with the symbol

$$\exp iD_x.D_\eta\, a(y,\eta)b(x,\xi)|y=x,\eta=\xi.$$

Note. The expansion of this symbol has the terms

$$(iD_\eta)^\alpha a_k(y,\eta)D_x b_k(x,\xi)/\alpha!|y= \ ,\eta=\xi.$$

Replacing a,b by a',b' multiplies this term by -1 to the power $j+k+|\alpha|$ which is at the same time its homogeneity. Hence (ab)'=a'b'.

Proof. The function b(x,D)u(x) has the Fourier transform

$$(bu)\hat{}(\eta)= (2\pi)^{-n}\int b(y,\eta)u\hat{}(\xi)\ \exp iy(\xi-\eta)\ dyd\xi,$$

and hence a(x,D)b(x,D) has the formal symbol

(3.1.2) $\quad (2\pi)^{-n}\int a(x,\eta)\exp ix.(\eta-\xi)\ d\eta\ \int b(y,\xi)\exp iy.(\xi-\eta)\ dy.$

Since the last integral is fast decreasing in $\xi-\eta$, this is really a symbol in x and ξ and this proves that the product is a pseudodifferential operator. The computations which follow reduce the expression above to normal form. With $\xi-\eta=\zeta$, $x-y=z$, we can write it as

$$(2\pi)^{-n}\ \int a(x,\xi-\zeta)d\zeta\ \int b(x-z,\xi)\exp ix.\zeta\ dz.$$

With $a^{(\alpha)}(x,\xi)=(iD_\xi)^\alpha a(x,\xi)$, let us develop $a(x,\xi-\zeta)$ in a Taylor series

$$\Sigma\ a^{(\alpha)}(x,\xi)(-\zeta)^\alpha/\alpha!\ +$$

$$+\quad \Sigma\ \int a^{(\alpha)}(x,\xi-t\zeta)(-\zeta)^\alpha t^{N-1}dt/\alpha!N!.$$

where the first sum runs over $|\alpha|<n$ and the second over $|\alpha|=n$. Inserting the first sum into (2) and using the properties of the Fourier transform gives a sum of terms

$$a^{(\alpha)}(x,\xi)(iD_x)^\alpha b(x,\xi)/\alpha!,\quad |\alpha|<N,$$

which are the first ones of our proposed asymptotic sum. The second term above when inserted into (2) gives rise to terms

$$\text{const}\int t^{N-1}dt\ \int a^{(\alpha)}(x,\xi-t\zeta)d\zeta\int(iD_x)^\alpha b(x-z,\xi)\exp ix.\zeta\ dz.$$

Here $a^{(\alpha)}$ has the majorant

$$\text{const } (1+|\xi-t\zeta|)^{m-|\alpha|},$$

where $m=m_a$ is the degree of a, and the last integral has the majorant

$$\text{const}((1+|\zeta|)^{-p}(1+|\xi|)^{m},$$

where $m=m_b$ is the degree of b. Taking p sufficiently large and separating the cases $t|\zeta|>|\xi|/2$ and $t|\zeta|<|\xi|/2$ in the integration results in the majorant

$$\text{const}((1+|\xi|)^{m-N}, \quad m=m_a+m_b,$$

for the terms of the second sum of (2). This finishes the proof.

Note. Principal symbols of adjoints, products and commutators.

It is easy to see that the class of polyhomogeneous operators is invariant under adjoints and products. An amplitude $a(x,\xi)$ for which there exists a non-zero homogeneous amplitude $a_m(x,\xi)$ of degree m such that $a- a_m$ has degree $<m$ is said to have the principal part a_m. It is obviously uniquely determined and it is the first term of the expansion of a when a is polyhomogeneous. When $P(x,\xi)$ is an amplitude with principal part $p(x,\xi)$, $p(x,D)$ is said to be the principal part of $P(x,D)$ and $p(x,\xi)$ its principal symbol. It follows from our theorems above that $p(x,D)$ with symbol $p(x,\xi)$ is the principal part of the adjoint of an operator whose principal symbol is $\bar{p}(x,\xi)$. Further, if P and Q are pseudodifferential operators with principal symbols $p(x,\xi)$ and $q(x,\xi)$, then $p(x,\xi)q(x,\xi)$ is the principal symbol of the product PQ and i times the Poisson commutator or parenthesis

$$\{p,q\}= p_x(x,\xi).q_\xi(x,\xi) -p_\xi(x,\xi).q_x(x,\xi),$$

where an index denotes the corresponding gradient, is the principal symbol of the commutator PQ-QP.

Note. Properly supported pseudodifferential operators.

Since the singular support of the kernel $K(x,y)$ of a pseudodifferential operator $a(x,D)$ in an open set X is contained in the diagonal $X \times X$, multiplying the kernel by any smooth function $f(x,y)$ which equals 1 in a neighborhood of the diagonal amounts to subtracting a smooth kernel. Hence, without changing its effect on singularities or its symbol we can replace any pseudodifferential operator by a properly supported one, i.e. an operator whose kernel has its support S so close to the diagonal that the sets of points (x,y) with x or y constant meet S in compact sets. It is clear that the product of two properly supported operators is again properly supported and that the symbols of the products of such operators can be computed according to the theorem above.

Pseudodifferential operators applied to oscillatory integrals

Sometimes one has to apply a pseudodifferential operator to an oscillatory integral. Under suitable restrictions, the result is another oscillatory integral with the same phase function and another amplitude. The precise result is as follows.

Let $s(x,\xi)$ be a regular phase function for x and ξ in R^n, let $a(x,\xi)$ be a polyhomogeneous phase function and let

$$u(x) = \int a(x,\xi) \exp is(x,\xi)\, d\xi$$

be the corresponding oscillatory integral.

Theorem Let $P(x,D)$ be a first order pseudodifferential operator with polyhomogeneous symbbol $P(x,\xi)$. If $a(x,\xi)$ vanishes for large enough x, then, modulo smooth functions, $P(x,D)u(x)$ is an oscillatory integral

$$\int b(x,\xi) \exp is(x,\xi)\, d\xi$$

where

(3.1.3) $b(x,\xi) \sim \sum P^{(\alpha)}(x,s_x(x,\xi)(D_x+r_x(x,y,\xi)^{\alpha}a(x\ \xi))|y=x$,

with

$$r(x,y,\xi) = s(x,\xi)-s(y,\xi) - s_x(x,\xi).(x-y).$$

Note. Since $r_y=0$ when $x=y$, the coefficients of $(D_y + ir_y)^{\alpha}$ have order at most $[\alpha/2]+1$ in ξ. Hence the formula for b gives a polyhomogeneous expansion.

Note. If the oscillatory integral $u'(x)$ is obtained from $u(x)$ by changing a to a' and s to $-s$, we have $(Pu)'=P'u'$. In fact, consider a term in the expansion (3),

$$P^{(\alpha)}{}_j(x,s_x(x,\xi)f(x,y,\xi)D_y{}^\beta{}_k(y,\xi)$$

where f is homogeneous of degree l in r. To compute $(Pu)'$ we have to change s to $-s$ and multiply the term above by -1 to the power $j+|\alpha|+l+k$. To get $P'u'$, we change s to $-s$, change η in $P_k(x,\eta)$ to $-\eta$ and multiply the term above by -1 to the power $j+|\alpha|+k+l$ where the l comes from the fact that as s changes to $-s$, r changes to $-r$.

Proof. The Fourier transform of $u(x)$ is the oscillatory integral

$$u\^{}(\eta) =\int a(x,\xi) \exp i(s(x,\xi)-ix.\eta)\, dxd\xi.$$

Since

$$Pu(y)=(2\pi)^{-n} \int P(y,\eta)u\^{}(\eta) \exp iy.\eta\, d\eta,$$

$Pu(y)$ equals the oscillatory integral

$$(2\pi)^{-n} \int P(y,\eta)a(x,\xi) \exp i(y-x).\eta +is(x,\xi)\, dxd\eta d\xi.$$

The rules for computing wave front sets of oscillatory integrals shows that only the parts of the integral close to where $x=y$ and $\eta=s_x(x,\xi)$ contribute to the wave front set of Pu. This makes the following expansion of $P(y,\eta)$ in terms of $\eta-s_x(x,\xi)$ a natural starting point,

$$(3.1.4)\quad P(y,\eta) =\Sigma P^{(\alpha)}(y,s_x(x,\xi))(\eta-s_x(x,\xi))^\alpha/\alpha! +$$
$$\Sigma P_\alpha(y,\eta,s_x(x,\xi)(\eta-s_x(x,\xi)^\alpha$$

for some smooth $P^{(\alpha)}(y,\xi,\eta)$ of degree $1-|\alpha|$ in ξ. The first sum is supposed to run over $|\alpha|<N$, the second over $|\alpha|=N$. The first sum inserted into the integral above and with the exponential rearranged as follows,

$$\exp(i(y-x).\eta +is(y,\xi) + is_x(x,\xi).(x-y) +ir(y,x,\xi))$$

calls for integration by parts in x with the result that we get a sum of terms with $|\alpha|<N$,

$(2\pi)^{-n} \int E(x,x,\xi,\eta)\ P^{(\alpha)}(y,s_x(x,\xi)(D_x)^\alpha a(x,\xi)\exp ir(y,x,\xi)/\alpha!dxd\eta d\xi$

where

$$E = \exp i(y-x).\eta + is(x,\xi) - ir(x,x,\xi).$$

This can be rewritten as

$$(2\pi)^{-n}\int \exp is(x,\xi)\ d\xi \int F(x,y,\xi)\ \exp i(y-x).\eta\ d\eta$$

where F is the general term of the expansion (3). Hence an integration with respect to η gives the desired expansion except that it has to be shown that the second term of (4) behaves properly. This somewhat lengthy exercise is left to the reader.

Changes of variables. Pseudodifferential operators on manifolds.

Let $f:X\to Y$ be a smooth diffeomorphism form an open set X in R^n to another open set Y. We shall see that f induces a linear bijection between corresponding pseudodifferential operators.

Theorem The map f induces a map

$$a(x,D) \to a_f(y,D_y),\ y=f(x),$$

where the symbol of the operator a_f is given by the formula

$$a_f(y,\eta)= \exp iD_{y'}.D_\eta\ b(y,y',\eta)|y'=y,$$

with

$$b(y,y',\eta)= a(g(y),{}^tg'(y')^{-1}\eta)|\det g'(y')\ F(y,y')|.$$

Here g(y) is the inverse of f(x) and F(y,y') is an nxn matrix defined by

$$(g(y)-g(y')).F(y,y')\xi = (y-y').\eta.$$

for y and y' close to each other.

Note. The theorem shows that, p being the principal symbol of a, the principal symbol of the operator a_f is

$$p(x,{}^tf'(x)\eta),\ x=g(y).$$

Note. In the expansion of $b_k(y,y',\eta)$ in powers of y-y' the coefficients of powers of y-y' have the form

$$C(y,\eta) = A(y,\eta)\eta^{\alpha}$$

where A is homogeneous in η of degree $-k-|\alpha|$. An η-derivative of order m of such a function is a sum of terms

$$B(y,\eta) = (PA(y,\eta))Q\eta^{\alpha}$$

where P and Q are η-derivatives of orders n and m-n. Multiplying A by $(-1)^k$ and changing η to $-\eta$ changes $B(y,\eta)$ to

$$(-1)^{n-k+|\alpha|}B(y,-\eta).$$

This is also what one gets by changing η to $-\eta$ in B while at the same time multiplying by -1 to the homogeneity m-k of B. Hence, taking the dual and changing variables are commutative operations when applied to a polyhomogeneous pseudodifferential operator.

Proof. The operator a(x,D) has the kernel

$$K(x,x') = \int a(x,\xi) \exp i(x-x').\xi \, d\xi.$$

Hence the transported operator has the kernel

$$K(g(y),g(y')) \, |\det g'(y')|$$

which equals

$$\int a(g(y),\xi) \exp i(g(y)-g(y')).\xi \, |\det g'(y')| d\xi.$$

Let G(y,y') be a matrix defined by

$$((g(y)-g(y')).\xi = (y-y').G(y,y')\xi.$$

so that $G(y,y)={}^tg'(y)$ for y'=y. We need only consider the kernel for y and y' close to each other and therefore the inverse F(y,y') of G is well defined. Putting $G(y,y')\xi=\eta$ and substituting shows that $a_*(y,D_y)$ has the kernel

$$\int b(y,y',\eta)\exp i(y-y').\eta \, d\eta.$$

That this gives the desired kernel is clear from the section dealing with adjoints. This finishes the proof.

Manifolds

It is clear from our last theorem that it makes sense to talk about pseudodifferential operators on a smooth manifold M and that the principal symbols of such operators are functions on its cotangent bundle T*M. In fact, given a principal symbol $p(x,\xi)$, it becomes $q(y,\eta)=p(x,\xi)$ in terms of coordinates (y,η) for which $\xi.dx=\eta.dy$. Also the principal symbol $i\{p,q\}$ of the commutator of two operators with principal symbols p and q is a function on the cotangent bundle.

3.2 L^2 estimates. Regularity properties of solutions of pseudodifferential equations

The main result about the size of pseudodifferential operators is that those of order 0 map square integrable functions with compact supports into locally square integrable functions. This will be proved below as part of a more general result. For this we need two lemmas.

Lemma If $N>\max(n,p+n)$, then, with integration over R^n,

$$\int (1+|s-t|)^{-N}(1+|t|)^{p}dt = O((1+|s|)^{p})$$

for any real p.

Proof. First, let $p>0$. Integrating over $|t|<|s|/2$ we get the desired majorant straight away. Integrating over $|t|>|s|/2$, we have $|s-t|>|s|/2$ and it suffices to estimate the integral of

$$(1+|t|)^{p-N}dt$$

over $|t|>|s|/2$. Since $N-p>n$, the integral converges to a bounded function of s. When $p<0$ and $|t|<|s|/2$,

$$(1+|s-t|)^{-N} < \text{const}(1+|s-t|)^{N-p}(1+|s|)^{p}.$$

and an integration with respect to t gives the desired result. When

$|t|>|s|/2$ we can proceed as before.- Our next tool is

Schur's lemma. If $K(s,t)$ is integrable and

$$\int|K(s,t)|dt \lesssim A, \quad \int|K(s,t)|ds \lesssim B,$$

then

$$\int(|\int K(s,t)f(t)dt|^2 ds \lesssim AB \int|f(s)|^2\, ds.$$

Proof. The left side of the conclusion is majorized by

$$\int K(s,t)K(s,t')|(|f(t)|^2+|f(t')|^2)dsdtdt'/2$$

and the result follows.

Let H^s be the space of distributions u for which the norm square with integration over R^n,

$$\|u\|_s{}^2 = \int|u(\xi)|^2(1+|\xi|)^{2s}d\xi$$

is finite. Here s is any real number. When s increases, the corresponding space decreases. We get a scale of spaces H^s where those with a sufficiently large negative s contain any given distribution with compact support.

Theorem If $P(x,D)$ is a pseudodifferential operator of order m in R^n and f is smooth with compact support, then

$$\|fP(x,D)u\|_s \leq \text{const } \|u\|_{m+s}$$

for all s and all distributions u.

Proof. We have

$$v(x) = f(x)P(x,D)u(x) = (2\pi)^{-n}\int \exp ix.\xi\; f(x)P(x,\xi)u(\xi)d\xi.$$

Taking the Fourier transform of the left side gives

$$v\hat{}(\eta) = \int K(\xi-\eta,\xi)u\hat{}(\xi)d\xi,$$

where K is of fast decrease in its first variable and of degree m in the second one. Hence

$$(1+|\eta|)^s v\hat{}(\eta) = \int G(\eta,\xi)(1+|\xi|)^{m+s}u\hat{}(\xi)d\xi$$

where, for any $N>0$,

$$G(\eta,\xi) = O((1+|\eta|)^{s}(1+|\xi-\eta|)^{-N}(1+|\xi|)^{-s}).$$

By the first of our lemmas above, this kernel meets the requirements of the second one. Hence the desired result follows.

The wave front sets of solutions of pseudodifferential equations

We have seen that the wave front set of Pu is contained in that of u when P is a pseudodifferential operator and u is a distribution. Below we shall give this statement a quantitative form in terms of the regularity function and also give a partial converse of it containing the wellknown fact that if P is an elliptic differential operator, u a distribution and Pu=0 then u is a smooth function.

Subsets Y of $R^n X R^n$ are said to be conical if $(x,t\xi)$ is in Y when $t>1$ and (x,ξ) is in Y. It is said to be conically compact if the set of points $(x,\xi/|\xi|)$ with (x,ξ) in Y is compact. Two points are said to be conically close when one of them is contained in a conical neighborhood of the other.

Given a conical neighborhood U of a pair (y,η), consider pseudodifferential operators

$$T=T(U) = f(x)A(D),$$

where f and $A(\xi)$ are smooth and A is homogeneous of degree zero for large ξ, their product vanishes outside a conically compact subset U and $f(x)=A(\xi)=1$ when (x,ξ) is conically close to (y,η). In terms of such operators we can define the regularity function r_u of u at (y,η) as the least upper bound as U tends (y,η) of numbers s for which Tu belongs to H^s.

Similarly, if $P(x,\xi)$ is an amplitude, we define the degree $m_P(y,\eta)$ of P at (y,η) as the greatest lower bound as U tends to (y,η) of deg TP.

Theorem If P is a pseudodiffferential operator and u is a distribution, then

$$r_{Pu}(x,\xi) \geq r_u(x,\xi) - m_P(x,\xi)$$

for all x and ξ. There is equality with m at the last place if P has a principal part of degree m which does not vanish conically close to (x,ξ).

Note. If P is elliptic in the sense that it has a principal symbol which never vanishes, the requirement for equality is satisfied everywhere and we get the classical result that a distribution is a smooth function outside the support of Pu.

Proof. Let T=T(U) and S=T(V) be as above, let V be fixed and U so small that $S(x,\xi)=1$ on the support of $T(x,\xi)$. By the preceding theorem,

$$\|TPSu\|_s \leq \text{const}\|Su\|_{s+m},$$

where m is the degree of TP. Further, by the rule for computing the symbols of products, that of TP(1-S) is fast decresing. Hence the inequality above holds for TPu if we add a constant to the right side. By letting U and V tend to (y,η), it follows that $\|u\|_s$ is finite when s+m is less than but arbitrarily close to $r_u(y,\eta)$ with m larger but arbitrarily close to $m_P(y,\eta)$. This proves the first part of the theorem. To prove the second part, we shall need the following result.

Lemma . Suppose that an amplitude $a(x,\xi)$ has a principal part a_m of degree m which does not vanish in some conical neighborhood U of a point (y,η). Then there is an amplitude $b(x,\xi)$ with principal part $1/a_m$ close to (y,η) such that

$$a(x,D)b(x,D)-1$$

has an amplitude of fast decrease close to (y,η).

Proof. Let us denote by $a(x,\xi)\circ b(x,\xi)$ the symbol of $a(x,D)b(x,D)$ given by the rules for computing the symbols of the products of pseudodifferential operators. Keeping (x,ξ) in U we shall compute one after the other of the terms in an asymptotic series for $b(x,\xi)$ with terms b_{-m-j} of degree $-m-j$,

$$b_{-m}(x,\xi)+b_{-m-1}(x,\xi)+\ldots .$$

Letting B_j be the sum of the first $j+1$ terms of the series, assume that

$$a(x,\xi)\ B_j(x,\xi)=1+c_{j+1}(x,\xi)$$

where c_{j+1} has degree $-j-1$. For $j=0$, this is achieved by putting

$$b_m(x,\xi)=1/a_m(x,\xi).$$

a_m being the principal part of a. In the step from j to j+1, b_{-m-j-1} should solve the equation

$$a(x,\xi)\circ(B_j(x,\xi)+b_{-m-j-1}(x,\xi))=$$
$$1+c_{j+1}(x,\xi)b_{-j-m-1}(x,\xi)+R$$

where R has the degree $-j-2$. Putting

$$b_{-m-j-1}(x,\xi)=\ -c_{j+1}(x,\xi)/a_m(x,\xi)$$

achieves the step of the induction. Let $B(x,\xi)$ be an amplitude with the asymptotic expansion above and let $f(x,\xi)$ be an amplitude of degree 0 which equals 1 in a conical neighborhood of (y,η) contained in U. The amplitude $b(x,\xi)=f(x,\xi)B(x,\xi)$ meets the requirements of the lemma.

Proof of the second part of the theorem. By assumption, there is a fixed conical neighborhood W of (y,η) where the principal part of the symbol of P has fixed degree m and does not vanish. Hence, by the lemma, there is a pseudodifferential operator Q such that that the symbol $P\circ Q(x,\xi)$ equals 1 in W. By the proof of the first part of the theorem,

$$\|TQSu\|_s \lesssim \text{const}\|u\|_{s-p},$$

when the supports of T and S are sufficiently close to (y,η) and p is any number >m. Here we can replace u by Pu. By the rules for computing the symbols of products, the symbol of the commutator of P and S has degree $-\infty$ on the support of $T(x,\xi)$. Hence $T[P,S]u$ is a smooth function with compact support. Hence the inequality above holds with TSu on the left and a constant added on the right so that, finally,

$$r_u(y,\eta) \geq r_{Pu}(y,\eta) + m$$

and this finishes the proof of the theorem.

Pseudodifferential operators with real principal symbols of principal type. Propagation of singularities theorem.

When a distribution u solves the differential equation $D_1u=0$, i.e. when u is independent of x_1, we shall see that its regularity function is independent of x_1. In fact, to measure the regularity function over a point y, one looks at integrals

$$\int f(x)u(x) \exp ix.\xi \, dx$$

where f is smooth with support close to y and $f(y)\neq 0$. Putting $g(x)=f(x+y)$ changes the integral to

$$\exp iy.\xi \int \exp ix.\xi \, g(x)u(x+y)dx$$

where g is smooth with support close to 0 and $g(0)\neq 0$. Hence, when u is independent of x_1, a shift of y in the x_1 direction just changes the integral above by an oscillating factor with no influence on size. This proves our statement.

The simple facts of the situation above extend to much more general situations. Let $P(x,D)$ be a pseudodifferential operator of order m with principal symbol $p(x,\xi)$ which is real and homogeneous of degree m and of principal type in the sense that its ξ gradient $p_\xi(x,\xi)$ never vanishes (for non-vanishing ξ). This principal symbol, in the example above equal to ξ_1, gives rise to a non-degenerate Hamiltonian system of first order differential equations,

$$x_t = p_\xi(x,\xi), \quad \xi_t = -p_x(x,\xi)$$

for curves $t \to (x(t),\xi(t))$ in the cotangent bundle of R^n, called bicharacteristics of p and P (and characteristics of the first order non-linear partial differential equations $p(x,v_x)=$ const). It follows from the equations that the function $p(x,\xi)$ is constant on every connected bicharacteristic. Those for which p vanishes are called nul bicharacteristics. When $p=\xi_1$, the bicharacteristics are lines parallel to the x_1 axis with ξ constant. On the nul bicharacteristics, ξ_1 is zero. In the sequel we only consider connected bicharacteristics.

Theorem Let P be a pseudodifferential operator with a real principal symbol p of principal type and degree m. Then, outside the wave front set of Pu, the regularity function of u is constant along any nul bicharacteristic. If the regularity function of Pu is at least s at B and that of u is s+m-1 at a point of B, then the regularity function of u is at least s+m-1 everywhere on B.

Note. Outside the wave front set of Pu and on the non-nul bicharacteristics, the regularity function of u is infinite. This follows from the preceding theorem. In an obviuos sense the last statement of the theorem is a propagation of singularities result. It is the main result of this kind and it is due to Lars Hörmander (1970).

Proof. Since neither the data of the theorem nor its conclusion change if we multiply P by an elliptic operator, we may assume that the degree of p is 1.

Let T be the transport along bicharacteristics from $t=t_1$ to $t=t_2$ and let B be a nul bicharacteristic with end points (y,η) and $(Ty,T\eta)$. The idea of the proof is to construct pseudodifferential operators Q and F^2, G^2 of the same degree such that Q is supported close to B, F close to (y,η) and G close to $(Ty,T\eta)$, and such that, approximately,

(3.2.1) $\quad i[Q,R] = F^2 - G^2.$

Then, if P,F,G are selfadjoint and Pu=0, we have, still approximately,

$0 = i((PQu,u)-(RQu,u)) = i([P,Q]u,u) = \|Fu\|^2 - \|Gu\|^2.$

If F can be chosen more or less arbitrarily and the same for G, the conclusion must be that the regularity function of u is constant along nul bicharacteristics. We shall show in what follows that it suffices to satisfy (1) on the symbol level.

Let R be a pseudodifferential operator of degree s whose symbol has its support close to B. If Pu=v we get

$$-(Rv,Ru) = -(RPu,Ru) = ([P,R]u,Ru) + (PRu,Ru),$$

where each term has a sense and the equality is true when s is sufficiently large negative. We shall try to determine R using the Poisson parenthesis $\{p,r\}$, where r is the principal symbol of R, as a differential operator on r. To do this, let P=A+iB where both A and B are selfadjoint. Then A has the principal symbol p and the degree of B is at most 0 so that $|(BRu,Ru)| \leq b(Ru,Ru)$ for some constant b. Taking the imaginary part of the above and using Schwarz's inequality we get

(3.2.2) $\quad \operatorname{Im}(R^*[P,R]u,u) - (1+b)(Ru,Ru) \leq (Rv,Rv)$

At this point it is convenient to state and prove a lemma about the differential equation for R inherent in the left side.

Lemma. Suppose that $f(x,\xi)$ is homogeneous of degree s with its support S conically close to the t_1 end of B and let c be a real number. Then there exist an amplitude $g(x,\xi)$ of degree s supported in TS and a non-negative amplitude q of degree 2s supported on the bicharacteristics joining T and ST such that

$$\{p,q\} - cq = f^2 - g^2$$

Note. Observe that q, being nonnegative and smooth, has a square root r which is an amplitude with the same support.

Proof. In the formula above, $\{p,q\}$ is differentiation of q along a bicharacteristic. If $a(x,\xi)$ is a smooth function with $\{p,a\}=1$ close to B, a change of unknown functions $q \to q \exp -ca$, $f \to f \exp -ca/2$, $g \to g \exp -ca/2$ changes the equation above to a similar equation with c=0 in which case there exists a solution u with the required properties if we let g be f transported by T. This proves the lemma.

End of the proof of the theorem. If R has a real prinicpal part $r(x,\xi)$, the equation (2) amounts to

$$([p(x,D),r^2(x,D)]-cr^2(x,D)u,u) \leq \text{const}((Rv,Rv)+(R_0u,R_0u)),$$

where R_0 has degree s-1/2. In viw of the lemma, this gives

$$(3.2.2) \qquad (Fu,Fu)-(Gu,Gu) \leq \text{const}((Rv,Rv)+R_0u,R_0u))+\text{const}$$

where $F=f(x,D)$, $G=g(x,D)$ and where the last const depends on s. Suppose first that v is smooth close to B. Then, letting the support of f tend to (y,η) we get, noting that regularity functions are continuous from below,

$$r_u(y,\eta) \geq \min\ (r_u(Ty,T\eta),1/2 + \min r_u \text{ over } B).$$

Varying B over parts of itself, this proves that the regularity function of u is constant along B. If the regularity function of Pu is at least s at B and that of u is equal to s at a point of B, the formula (2) applied to a part of B near the point in question shows that the regularity function of u is at least s at all of B. This finishes the proof.

3.3 Lax's construction for Cauchy's problem and a hyperbolic first order differential operator

Lax's construction extends to hyperbolic first order pseudodifferential operators

$$P = D_t + Q(t,x,D_x)$$

and general Cauchy problems. Here Q is polyhomogeneous of degree 1,

$$Q(t,x,\xi) = \Sigma\ Q_k(t,x,\xi) \ , \ (k=1,0,\ldots),$$

with real principal part

$$q(t,x,\xi) = Q_1(t,x,\xi).$$

This means that the principal part $\tau+q(t,x,\xi)$ of P is real, in analogy with first order hyperbolic differential operators.

Theorem Let

$$v(x) = \int b(x,\eta) \exp i\, s(x,\eta)$$

be an oscillatory integral with a polyhomogeneous amplitude a of degree r. If the phase function $s(t,x,\eta)$ solves the Hamilton-Jacobi equation,

$$s_t + q(t,x,\xi) = 0, \quad s(0,x,\eta) = s(x,\eta),$$

there exists an oscillatory integral

$$u(t,x) = \int a(t,x,\eta) \exp is(t,x,\eta)d\eta$$

with a polyhomogeneous amplitude a of degree r which solves the Cauchy problem

$$Pu = \text{smooth}, \quad u-v = \text{smooth when } t=0$$

for small t and x.

Note. With $s(x,\eta)=x.\eta$, $r=0$ and $a(x,\eta) = 1/(2\pi)^n$, $v(x)=\delta(x)$ and $v(t,x)=H(t)u(t,x)$ with $H(t)=0$ when $t<0$ and 1 otherwise, the distribution $Pv-\delta$ is smooth and we get a fundamental solution of P with pole at 0.

Proof. Let us expand $a(t,x,\eta)$ in a series with in homogeneous terms,

$$a = \Sigma\, a_k, \quad (k=r,r-1,\dots).$$

By the rules for applying a pseudodifferential operator to an oscillatory integral, Pu is an oscillatory integral with phase function s and amplitude

$$s_t a + s a_t + \Sigma\, Q^{(\alpha)}(t,x,s_x(t,x,\xi)(D_x + r_x(t,x,y,\ )^\alpha a(t,x\ \xi)/\alpha!|y=x$$

where

$$r(t,x,y,\xi) = s(t,x,\xi) - s_y(t,x,\xi) - s_x(t,x,\xi).(x-y).$$

The term of degree r+1 is

$$(s_t + q(t,s,s_x))a_r$$

and vanishes by assumption. The term of degree r is La_r where

$$L = \partial_t + a_j(t,x,s_x).iD_x + Q_0$$

is a linear differential operator of degree 1. The equation $La_r=0$ has a unique solution a_0 when a_0 is fixed for t=0. We give it the value $b_r(x,\eta)$. The terms of degree k involve the parts $a_0,\dots,a_k$ of a and has the form

$$La_k + M$$

where M does not involve a_k. Putting $a_k = b_k$ for all k and $t=0$ fixes all a_k with $k>0$. The result is a sequence of terms $a_0, \ldots$ such that their sum with the terms suitable adjusted at $\xi=0$ is a polyhomogeneous amplitude a of degree 0 which fulfils the requirements of the theorem. The proof is finished.

CHAPTER 4

THE HAMILTON-JACOBI EQUATION AND SYMPLECTIC GEOMETRY

We have already met the Hamilton-Jacobi equation in Lax's construction of a local parametrix for the fundamental solution of a hyperbolic equation. The construction of global parametrices requires a fair amount of differential geometry connected with this equation, in particular the simplest facts about Lagrangian manifolds. This material will be reviewed below.

Hamilton systems

A Hamilton system of ordinary equations is one of the form

(4.1.1) $\quad x_t=H_p(x,p,t),\ p_t=-H_x(t,x,p)$

where $x=(x_1,\dots,x_n)$ and $p=(p_1,\dots,p_n)$ are functions of t and H is a given smooth real function of t,x,p. The indices indicate the corresponding gradients. By the local theory of such systems, the Hamilton flow

$$T:(s,y,q)\to(t,x,p)\quad (x,p=y,q \text{ when } t=s)$$

is a local smooth bijection for fixed s and t. The theory of Hamilton systems depend on the following

Theorem The differential of the differential form

$$w=p.dx-Hdt,\quad (p.dx=p_1dx_1+\dots+p_ndx_n)$$

is invariant under the Hamilton flow, i.e. $dw(t,x(t),p(t)=dw(s,y,q)$. When H is homogeneous of degree 1 in p, w itself is invariant.

Proof. For simplicity, let s=0. Using an obvious notation we have

$$dx=x_ydy+x_qdq+x_tdt,\ dp=p_ydy+p_qdq+p_tdt,$$

$$dH=H_xdx+H_pdp+H_tdt.$$

Hence, if $d\bar{x}$ and $d\bar{p}$ denote dx and dp with dt=0, we get

$$dw=dp\wedge dx-dH\wedge dt = d\bar{p}\wedge d\bar{x} +p_t dt\wedge d\bar{x}+d\bar{p}\wedge x_t dt-$$
$$-H_x d\bar{x}\wedge dt-H_p d\bar{p}\wedge dt.$$

Here everything involving dt disappears due to the equations (1). It follows that dw is a linear combination of the differentials $dy\wedge dy$, $dy\wedge dq$ and $dq\wedge dq$ with coefficients depending, maybe, on t. But since ddw=0, all the t-derivatives of the coefficients of dw vanish and hence dw is independent of t. When H is homogeneous of degree 1, $p.H_p=H$ so that

$H_x dx+H_p dp=dH=H_p dp+pdH_p$. Then $H_x dx=pdH_p$ and hence, with differentiations along the flow,

$$d(p.x)/dt=p_t dx+pdx_t=-H_x dx+pdH_p=0.$$

This finishes the proof.

Our theorem can be used to find more or less explicit solutions of Hamilton-Jacobi's equation

(4.1.2) $\quad f_t+H(t,x,f_x)=0,\ f(0,x)=g(x)$

with a given g(x). In fact, w is closed on every manifold Y of the form t=0, q=g'(y) with y in some open bounded part of R_n, and hence also on the Hamilton outflow T(Y) of Y. When t is small enough, T(Y) meets the hyperplanes t=const in manifolds Y(t) where we can choose x=Y(t) as coordinates. In terms of these coordinates, w=pdx, x=x(t), p=p(t) so that w=df(t,x) for some smooth function f(t,x) with

$$f_t(t,x)=-H(t,x,p),\ f_x(t,x)=p,$$

were x=x(t), p=p(t). It follows that f solves the Hamilton-Jacibi equation (2). By general theory, the solution is unique. Along the flow we have

$$(d/dt)f(t,x)=f_t+f_x dx = -H+pH_p$$

so that

(4.1.3) $f(t,x(t)=g(x(0))+ \int_0^t (p.H_p-H)(s,x(s),p(s))ds.$

Hence, if H(t,x,p) is homogeneous of degree 1 in P, f is obtained from its intial value g(x) by propagation along the flow. The formula ceases to hold when the flow from Y no longer has a surjective projection on x-space. We formulate the results in a theorem.

Theorem Let A be a bounded part of R^n, g a smooth real function form A and Y the manifold t=0, q=g'(y), y in A. Then the differential form w=p.dx-H(t,p,x)dt is closed on the Hamilton ouflow T(Y) from Y. When the number a>0 is so small that the maps y->x(t,Y) are smooth bijections when |t|<a, w is the differential of a functions s(t,x) satisfying the Hamilton-Jacobi equation (2. It is given explicitly by (3) and it is constant along the flow if H(t,x,p) is homogeneous of degree 1 in p.

In this theorem, the differential form is the global object while the solution of the Hamilton-Jacobi equation is a local one, We shall se later that also the solution s is a global one in a certain sense. To do this we shall need some geometric prerequisites given in the next section.

4.2 Symplectic spaces and Lagrangian planes

A linear space S of dimension 2n is said to be symplectic when equipped by a non-degenerate skewsymmetric form W(X,Y) which is non-degenerate in the sense that W(X,S)=0=>X=0.

Theorem The space S has bases, called canonical, with elements $e_1,...,e_n$, $\epsilon_1,...,\epsilon_n$ such that

$$W(e_j,e_k)=0, W(\epsilon_j \epsilon_k)=0, \quad W(e_j,\epsilon_k)=\delta_{jk}.$$

Proof. Since the form is non-degenerate, every X≠0 has some companion Y such that W(X,Y)=1. It follows from this formula that X and Y are linearly independent. Hence there are elements e_1 and ϵ_1 which satisfy the formulas above when j,k=1. Since the subspace of S for which w vanishes on the span of e_1 and ϵ_1 is again a symplectic space of dimension 2n-2 on which w is a non-degenerate, the construction can be repeated till we get a nul space. This completes the proof.

A linear bijection $U:S\to S$ is said to be symplectic if it preserves W, i.e.

$$W(X,Y)=W(UX,UY)$$

for all X and Y. For this it is necessary and sufficient that U maps a canonical basis into a canonical basis. In terms of the matrix of U relative to a canonical basis,

$$Ue_j=\Sigma a_{jk}e_j + \Sigma b_{jk}\epsilon_k,$$

$$U\epsilon_j=\Sigma c_{jk}e_k + \Sigma d_{jk}\epsilon_k,$$

this means that

$$AC'=0, B'=0, AD'-BC'=I$$

for the corresponding matrices. The symplectic structure of S can be transferred to its dual S^* consisting of linear forms F on S. When F and G are two such forms, we define $F\wedge G$ as a bilinear skewsymmetric form on SXS defined by

$$(F\wedge G)(X,Y)=F(X)G(Y)-F(Y)G(X)$$

for all X and Y in S. With this notation we have

(4.2.1) $W=f_1\wedge\varphi_1+\ldots+f_n\wedge\varphi_n,$

where

$$f_1,\ldots,f_n,\varphi_1,\ldots,\varphi_n$$

is a basis of S^* biorthogonal to a canonical basis $e_1,\ldots,\epsilon_n$ of S in such a way that f_1 maps all the basis elements to zero except that e_1 is mapped to 1 etc. In fact, if

$$X= \Sigma a_ie_i + \Sigma b_i\epsilon_i,$$

$$Y= \Sigma c_ie_i + \Sigma d_i\epsilon_i,$$

it simply expresses the fact that

$$W(X,Y) = \Sigma a_id_i -\Sigma b_ic_i$$

Putting

$$UF(X)=F(UX),$$

every linear map of S induces one of S^*. The formula (1) above shows that U is symplectic if and only of U, acting on S^*, preserves the right side of the formula.

Lagrangian planes

Linear subspaces of a symplectic space where W vanishes identically are said to be isotropic. They must have dimension $\leq n$. In fact, if L is such a subspace of dimension $p>n$ and M is a complement in S, then $W(f,M)=0$ with f in L amounts to $2n-p$ linear equations in p variables and the existence of a solution f implies that $W(f,S)=0$, i.e. $f=0$. But if $p>n$, then $2n-p<p$ and a non-zero solution exists which is a contradiction. Hence $p\leq n$.

Isotropic subspaces of maximal dimension are said to be Lagrangian.

Examples. Any plane spanned by one half of the members of a canonical basis of S is Lagrangian. It is not difficult to show that every Lagrangian plane has this form when referred to a suitable canonical basis.

Parametrization of Lagrangian planes

Let C^n be complex n-space with basis $e_1,\dots,e_n$ and consider a unitary metric Q of the form

$$Q(z,w)= \Sigma z_k\bar{w}_k$$

where $z=z_1e_1+\dots$ and the same for w.

Lemma Write $z=x+i\xi$. $w=y+i\eta$. Then

$$W(z,w) = \operatorname{Im} Q(z,w)$$

is a bilinear non-degenerate symplectic form for which

$$e_1,\dots,en,\epsilon_1,\dots,\epsilon_n$$

is a canonical basis.

Proof. The supposed basis is linearly independent over the reals and, since $W(z,w)=\xi.y-\eta.x$, they form a canonical basis.

Lemma The image of $R^n \subset C^n$ under a unitary map is a Lagrangian plane and every Lagrangian plane can be obtained in this way.

Proof. That U is unitary means that $Q(Uz,Uw)=Q(z,w)$ for all z and w. If z and w are real, the right side is real so that $W(Uz,Uw)=0$. Hence UR^n is a Lagrangian plane. Conversely, let L be a Lagrangian plane. It is the real span of vectors $f_1,\dots,f_n$ such that $W(f_j,f_k)=0$ for all j and k. Since $\mathrm{Re}\, Q(z,w)=x.y+\xi.\eta$ defines a definite metric, the f_j can be made to satisfy $\mathrm{Re}\, Q(f_j,f_k)=\delta_{jk}$. But then $f_1,\dots$ are obtained from $e_1,\dots$ by a unitary map. This proves the lemma.

Note. Since $UR^n=R^n$ if and only if U is orthogonal, the space of Lagrangian planes is homeomorphic to the quotient $U(n)/O(n)$.

When $z=Uw$ we have

$$x= \mathrm{Re}\, Uy - \mathrm{Im}\, U\eta,\quad \xi = \mathrm{Im}\, Uy + \mathrm{Re}\, U\eta.$$

Here Re U need not be invertible. It is important for us that Re U can be made invertible by a change of variables

$$y=y',\quad \eta= \eta'+ty$$

which is symplectic. In fact, then

$$x=(\mathrm{Re}\, U +t\, \mathrm{Im}\, U)y',\quad \xi = (\mathrm{Im}\, U +t\, \mathrm{Re}\, U)y' + \mathrm{Re}\, U\, \eta'.$$

If $\det(\mathrm{Re}\, U +t\mathrm{Im}\, U)=0$ for all t, so also for $t=-i$, i.e. $\det U=0$, a contradiction. Hence $\mathrm{Re}\, U - t\mathrm{Im}\, U$ is invertible except when t is a zero of a certain polynomial.

4.3 Lagrangian submanifolds of the cotangent bundle of a manifold

Let X be a smooth manifold of dimension n. The cotangent space over a point p of X is spanned by the differentials df of smooth real functions from X restricted to p. If $x=(x_1,\dots,x_n)$ are coordinates at p with p corresponding to 0, the cotangent space over p is the linear span of the coordinate differentials

$$w=\xi_1 x_1+\ldots+\xi_n x_n$$

restricted to p. Here $\xi=(\xi_1,\ldots,\xi_n)$ are coordinates in the cotangent plane. The dual of the cotangent plane is called the tangent plane over p. It consists of linear forms in the coordinates ξ and can also be thought of as the span of the differentiations $\partial/\partial x_k$ at x=0. This is the geometric meaning of the tangent plane at p. The cotangent plane can also be thought of as the space of linear forms on the tangent plane.

Combining a manifold X and all its cotangent planes we get a new manifold, the cotangent bundle $T^*(X)$ of (or over) the base X. Locally, it has coordinates x and ξ and appears as the product of an open subset of R^n and a linear space. It gets its structure by the requirement that the differential form w be invariant, i.e. that

$$\xi.dx=\eta.dy$$

when the coordinates x,ξ and y,η overlap. We shall refer to w as the canonical 1-form. Since it is invariant, its differential

$$dw=d\xi\wedge dx = \Sigma\ d\xi_k\wedge dx_k$$

is also invariant which means that the cotangent bundle has a symplectic structure and that everyone of its cotangent planes or tangent planes is a symplectic space.

Like any manifold, the cotangent bundel has submanifolds, for instance its sections, sunmanifolds given locally by $\xi=f(x)$ for some smooth f. When f=0, this definition is global and defines X as a submanifold of the cotangent bundle called its zero section.

In a certain sense opposite to the sections are the Lagrangian submanifolds of the cotangent bundle. They are submanifolds of dimension n on which the canonical form $\xi.dx$ and hence also its differential vanish. It follows that their tangent planes are Lagrangian planes in the tangent planes of the cotangent bundle. A Lagrangian manifold L is said to be homogeneous of $(x,t\xi)$ is in L when

(x,ξ) is and $t>0$.

Examples. Any submanifold of the cotangent bundle locally of the form

$$x=H_\xi(\xi)$$

where H is homogeneous of degree 1 is a homogeneous Lagrangian manifold. In fact, its dimension is n and $\xi.dx=\xi H_{\xi\xi}dx=0$ since H_ξ is homogeneous of degree 0. The word local has to be interpreted so that x stays close to some x_o and ξ conically close to some $\xi_o\neq 0$ in the sense that ξ is not zero and restricted to an open cone around ξ_o. We shall refer to this situation as x,ξ being conically close to x_o,ξ_o.
Note. The zero section of the cotangent bundle is obviously a homogeneous Lagrangian manifold. In the sequel we shall be interested in Lagrangian manifolds outside the zero section.

Before proving that our last example is general we need a

Lemma Let L be a homogeneous Lagrangian submanifold of $T^*(X)$ and suppose that the differentials $d\xi$ are linearly independent when restricted to a tangent plane T(L) of L at a point p with coordinates x_o,ξ_o. Then L has the form $x=H_\xi(\xi)$ conically close to p with a unique function H homogeneous of degree 1.

Proof. By hypothesis, there are smooth functions h_j such that $x_j=h_j$ on L close to p. Inserting this into $d\xi\wedge dx=0$ shows that $(h_1,...,h_n)$ is the gradient of a function $H(\xi)$ and the condition that $\xi.dx=0$ shows that $\xi.H_{\xi\xi}=0$, i.e. H_ξ is homogeneous of degree zero an then H is unique if chosen homogeneous of degree 1.

We shall now see that our example is general outside the zero section of the cotangent bundle.

Lemma Let L be a homogeneous Lagrangian manifold L of the cotangent bundle $T^*(X)$. Under any point p of L outside the zero section there are coordinates x on X and a function $H(\xi)$, homogeneous of degree 1 in the corresponding coordinates ξ in the fiber over p such that L is given by $x=H_\xi(\xi)$ conically close to p.

Proof. By an affine change of coordinates on X under p we can achieve that p is given by $y=0$, $\eta=(1,0,\ldots,0)$. Since the tangent plane of $T^*(X)$ at p is linear of dimension 2n, we can find a k such that the restrictions to T(L) of the differentials

$$d\eta_1,\ldots,d\eta_k,dy_{k+1},\ldots,dy_n$$

are linearly independent. Now make a change of variables,

$$x_1=y_1+z,\quad x_2=y_2 \text{ etc}$$

where

$$z=(y_{k+1}^2+\ldots+y_n^2)/2.$$

The rule $\eta.dy=\xi.dx$ then gives the new ξ coordinates

$$\eta_j=\xi_j \text{ when } j\le k,\quad \eta_j=\xi_j+\xi_1 y_j \text{ when } j>k.$$

Hence, at p we have

$$d\xi_j=d\eta_j \text{ when } j\le k,\quad d\xi_j=d\eta_j-dy_j \text{ when } j>k.$$

It follows that $d\xi_1,\ldots,d\xi_n$ are linearly independent on T(L). Hence an appeal to the previous lemma finishes the proof.

4.4 Hamilton flows on the cotangent bundle. Very regular phase functions

Let $H(t,x,\xi)$ be a smooth real function, homogeneous of degree 1 in ξ, on the product of a real line and the cotangent bundle of a manifold X of dimension n. The corresponding Hamilton system,

$$x_t=H_\xi(t,x,\xi),\quad \xi_t=-H_x(t,x,\xi)$$

is then invariant, i.e. the form of the equation does not change under coordinate transformations leaving $\xi.dx$ invariant. The verification of this is immediate. Hence the corresponding Hamilton flow

$$(x,\xi)(s)\to(x,\xi)(t)$$

is globally defined on $T^*(X)$. Since H is homogeneous of degree 1 in ξ, the flow commutes with maps $\xi\to\text{const }\xi$ and hence preserves the homogeneity structure of cotangent planes. By a previous result, the canonical differential form $\xi.dx$ is invariant under the Hamilton flow. Finally, let us note that an orbit from a point where ξ vanishes remains at that point. Hence the Hamilton flow from a Lagrangian manifold outside the zero section of the cotangent bundle remains outside for all time.

Since the canonical form $\xi.dx$ is invariant under the Hamilton flow, it maps a Lagrangian manifold L given for $t=0$ into a family of Lagrangian manifolds $L(t)$. We shall look into the details of this map.

Very regular phase functions along a bicharacteristic tube

In the construction of global parametrices for hyperbolic pseudodiffeential equations in the next chapter, the following definition is crucial.

Definition Canonical coordinates (x,ξ) near a point p in $T^*(X)$ are said to be regular for a homogeneous Lagrangian manifold L through p if p has a neighborhood where the variables ξ can be taken as parameters on L.

Arbitrary canonical coordinates (x,ξ) at p can be made regular locally for L by a change of variables. This follows from the Theorem of section 4.2. We can now define very regular phase functions.

Definition A phase function $g(x,\xi)$ is said to be very regular for L at p if the equations $g_\xi=0$ define x as a function of ξ close to p in such a way that $g_\xi=0\Rightarrow dg=\xi.dx$ and $(x,\xi=g_x)$, $g_\xi=0$ parametrices L close to p.

Example. A phase function $x.\xi-H(\xi)$ with H homogeneous of degree 1 is very regular for the homoeneous Lagrangian manifold $x=H'(\xi)$. Hence every homogeneous Lagrangian manifold has a very regular phase function locally.

Note. The notion of very regular phase function is invariant under changes of the x variables and concomitant changes of the ξ variables. In fact, if x' are the new variables, $x=x(x')$ and $\xi'.dx'=\xi.dx$ and $\partial\xi/\partial\xi'$ is not degenerate, then g_ξ and $g'_{\xi'}=g_\xi\,\partial\xi/\partial\xi'$ vanish at the same time and when they vanish, then $dg=\xi.dx=\xi'.dx'=dg'$. This observation gives some plenty of examples of very regular phase functions.

Very regular phase functions under the Hamilton flow

Let B be a nul bicharacteristic for H, i.e. the trace of the Hamilton flow starting from a point q for $t=0$ to a point p for $t=s$. Let (y,η) be canonical coordinates for $L(0)$ around $q=(y_0,\eta_0)$ and x,ξ coordinates around $p=(x_0,\xi_0)$. The flow maps a conical neighborhood $N(0)$ of q to a conical neighborhood $N(s)$ of p and the outflow of $N(0)$ is a tube of bicharacteristics around B. A Lagrangian manifold $L(0)$ of $N(0)$ containing q is mapped to a Lagrangian manifold $L(s)$ containing p.

Theorem It is possible to introduce regular canonical coordinates (y,η) at q and regular canonical coordinates (x,ξ) at p such that, if $g(y,\eta)$ is a very regular phase function for $L(0)$ in $N(0)$, the function
$f(s,x,\xi)=g(y,\eta)$
where (y,η) and (x,ξ) are connected by the Hamilton flow, is a very regular phase function for $L(s)$ at p.

Proof. Assume that the choice of coordinates has been made so that the variables ξ can be chosen as parameters on $L(s)$. The map $(y,\eta)\to(x,\xi)$ is invertible and at p,q we have

$$dx=x_y dy+x_\eta d\eta,\quad d\xi=\xi_x dx+\xi_\eta d\eta.$$

We should like to have the implication $dx,d\eta=0 \Leftrightarrow dy,d\eta=0$, i.e. x_y

should be non-singular. This can be achieved by a change of coordinates

$$y_1'=y_1+\epsilon z,\quad y_2'=y_2 \text{ etc}$$

where ϵ is small and

$$z=(y_1^2+...+y_n^2)/2$$

In fact, then $d\eta=d\eta'+\epsilon dy'$ so that x_y changes to $x_y+\epsilon x_\eta$. As remarked at the end of section 4.2, there are arbitrarily small ϵ for which $x_y+\epsilon x_\eta$ is non-singular. Hence the desired implication can be achieved by a change of variables and when it is satisfied, we can define a function $h(x,\eta)$ by

$$h(x,\eta)=g(y,\eta).$$

Then

$$dh=h_x dx+h_\eta d\eta=g_y dy+g_\eta d\eta.$$

On $L(0)$, g_η vanishes and $g_y=\eta$ so that the right side equals $\eta dy=\xi.dx$. Hence $h_\eta d\eta=0$ for all $d\eta$ which means that $h_\eta=0$. Hence

$$(x,h_x),\quad h_\eta=0$$

parametrices $L(s)$ close to p. Since $(x,\eta)\to(x,\xi)$ is a bijection (via the bijection $(y,\eta)\to(x,\xi)$), we have $f(s,x,\xi)=h(x,\eta)$ close to p. On $L(s)$ we have $h_\eta=0$ which means that $df=h_x dx$. Hence

$$(x,\xi=f_x),\quad f_\xi=0$$

parametrices $L(s)$, i.e. f is a very regular phase function for $L(s)$. The proof is finished.

References

All the material of this chapter except the notion of very regular phase function is standard (see Hörmander 1985, III Ch. XXI). Invariant differential forms were introduced by Poincare and extensively used by Elie Cartan.

CHAPTER 5

A GLOBAL PARAMETRIX FOR THE FUNDAMENTAL SOLUTION OF A FIRST ORDER HYPERBOLIC PSEUDODIFFERENTIAL OPERATOR.

Introduction The motives for the study of Lagrangian manifolds and Fourier integral operators have been the semiclassical approximations to quantum physics (Maslov 1965) and the construction of global parametrices of the fundamental solutions of hyperbolic differential equations with variable coefficients (Duistermaat and Hörmander 1972). In this chapter we shall give a simple construction of such a parametrix for a first order hyperbolic pseudodifferential operator. It illustrates most of the difficulties and the techniques of the field.

5.1 Cauchy's problem for a first order hyperbolic pseudodifferential operator

A first order pseudodifferential operator in the variables t and $x=(x_1,\dots,x_n)$ is said to be hyperbolic with respect to t if it has the form

$$P = D_t+Q(t,x,D)$$

where the principal symbol $q(t,x,\xi)$ of Q is real of degree 1 and the complete symbol $Q(t,x,\xi)$ is supposed to be polyhomogeneous. Such operators appear in Lax's construction of a parametrices for strongly hyperbolic differential operators. They are essential generalizations of the corresponding first order differential operators. As an example of how to handle pseudodifferential operators we shall solve the following Cauchy problem

(5.1.1) $\quad Pu=0$ when $t>0$, $u=w$ when $t=0$

and construct a parametrix for the corresponding fundamental solution.

To simplify we shall assume that Q is special in the sense that the symbol $Q(t,x,\xi)$ vanishes when $|x|$ exceeds some continuous function of t. In what follows, $u(t)$ denotes $u(t,x)$ as a distribution on R^n.

Theorem Let s be any real number. If $v(t)$ is integrable as a function with values in H^s and w is in H^s, Cauchy's problem (1) has a unique weak solution u such that $u(t)$ is continuous with values in H^s. In addition, if I is the interval from 0 to T,

(5.1.2) $$\|u(T)\|_s < \exp cT(\|w\|_s + \int_I \|v\|_s \, dt)$$

for all s and T where c is a locally bounded function of these variables.

Note. The theorem also applies to (1) in the region $t<0$ and a distribution which is a solution for $t>0$ and $t<0$ solves $Pu=0$ for all t.

Proof. We note first that since

$$Qu(t,x) = (2\pi)^{-n} \int \exp ix.\xi \; Q(x,t,\xi) \; \hat{u}(\xi) d\xi,$$

Q operates from tempered distributions in x to distributions with compact supports. More precisely, by the continuity properties of pseudodifferential operators, Q maps H^s continuously into H^{s-1} and, more generally, functions $u(t)$ which are differentiable, continuous, integrable or essentially bounded with values in H^s to the same kind of functions with values in H^{s-1}. Since H^{-s} is the dual of H^s with respect to the duality

$$(u,v) = \int u(x)\bar{v}(x) dx,$$

the adjoint Q^* of Q has the same properties.

The main ingredient in the proof is the energy inequality (2). We prove it first under the assumption that $u(t)$ is continuously differentiable with values in H^{s+1}. Put

$$A(D) = 1+(D_1^2+...+D_n^2)^{1/2}$$

so that

$$\|u\|_s^2 = (A^s u, A^s u)$$

where the right side is the usual L^2 norm square over R^n. With u as above we then have

$$(d/dt)A^s u(t) = -iA^s Qu(t) + iA^s v(t), \quad v=Pu.$$

Hence

$$(d/dt)\ (u(t),u(t))_s = -i(Bu(t),u(t)) + 2\ \mathrm{Im}\ (u,v)_s.$$

where

$$B = A^{2s}Q - Q^*A^{2s}.$$

Since Q and Q^* have the same principal part, the order of B is 2s and this proves that

$$(d/dt)\|u(t)\|_s \le c\ \|u(t)\|_s + \|v(t)\|_s$$

with c bounded for t and s bounded. Hence (1) follows in this case. By a passage to the limit it also follows when the derivative of u is an integrable function with values in H^s. We note that, by virtue of its proof, the energy inequality holds also for the adjoint of P.

To prove existence, we shall see that the image of the map

$$u \to u(0), Pu(t)$$

with u continuously differentiable and with values in H^{s+1} is dense in the direct sum

$$H^s \oplus L^1$$

where the last term refers to functions of t with values in H^s which are integrable in some interval $I=(0,T)$. The dual of this direct sum is

$$H^{-s} \oplus L^\infty$$

where L^∞ refers to functions from I to H^{-s} which are essentially bounded. Hence it suffices to show that if w+v belongs to this sum and

(4.1.3) $$(u(0),w) + \int_I (Pu(t),v(t))\ dt$$

vanishes for all u above, then w and v vanish. First, we shall let u vanish near 0. This means that v(t) is a weak solution of the equation

$$(D_t + Q^*)v(t)=0.$$

Here Q^*v is integrable with values in H^{-s-1} and it follows that the derivative of v is an integrable function with values in H^{-m-1}. But then we can integrate by parts in (3) getting

$$0 = (u(T),v(T)) + \int_I (u(t),P^*v(t)).$$

Since u(T) is arbitrary in H^m and this space is dense in the dual H^{m+1} of H^{-m-1}, it follows that v(T) =0. But then the energy inequality applies to v and the interval I and shows that v=0. Then, finally, if we lift the restriction that u(0)=0, it follows from (2) that w=0. Hence our density statement is proved and with that, by a passage to the limit in (1), also the theorem.

5.2 Cauchy's problem on the product of a line and a manifold

Let X be a n-dimensional orientable manifold and P(t,x,D) a pseudodifferential operator of degree 1, polyhomogeneous and with principal symbol $q(t,x,\xi)$, defined on the product Y = X x R. Consider the following Cauchy problem for $P=D_t+Q(t,x,D)$ on Y,

(5.2.1) $\qquad$ Pu=v when t>0, u=w when t=0.

According to Hörmander's propagation of singularities theorem, the wave front set of a solution u is contained in the set of nul bicharacteristics

(5.2.2) $\quad x_t = q_\xi(t,x,\xi), \quad \xi_t = -q_x(t,x,\xi), \quad \tau=-q(t,x,\xi),$

issuing from the wave front sets of v and w. The Cauchy problem (1) cannot be solved in general even for differential operators. One of the difficulties occurs when the nul bicharacteristics approach the boundary of X. This can be avoided by the assumption that all Hamilton maps $x(a),\xi(a) \to x(b),\xi(b)$ send compact subsets of $T^*(X)$ into compact sets for all a and b. A difficulty for pseudodifferential operators is the existence of Pu when X is not compact. However, if Cauchy's problem (1) is taken modulo smooth functions we can assert that there is a unique soluton for all times and all distribution data v and w. The proof, which will not be given in detail, uses partitions of unity and small steps in time. In this way the problem is reduced to cases when the supports of v and w are contained in coordinate neighborhoods and the previous theorem can be used.

In order to solve (2) modulo smooth functions, it is sufficient to know fundamental solutions E(t,s,x,y) of P with the property that

$$P(t,x,D_t,D_x)E(t,x,s,y) = \delta(t-s)\delta(x,y), \quad E=0 \text{ when } t<s.$$

where $\delta(x,y)$ is a distribution such that

$$\int f(x)\delta(x,y)\omega(y) = f(y).$$

with ω some fixed smooth positive n-form on X. In particular, the distribution

$$u(t,x)= \int E(t,0,x,y)w(y)\omega(y)$$

solves the Cauchy problem above with v=0. The distribution

(5.2.3) $\qquad t,x \to F(t,x,y) = E(t,0,x,y).$

solves the same problem with $w=\delta(x,y)$. The wave front set of F is contained in the Hamilton outflow from the Lagrangian manifold $L(0)=((0,y),R^n\setminus 0)$. The image of L(0) at time t will be denoted by L(t). To get the complete wave front set of F considered as a function of both t and x, the pair $(t,\tau=-q(t,x,\xi))$ should be joined to L(t).

For small t, Lax's construction provides a parametrix of F (see section 3.3). In the next section we shall construct a global parametrix.

5.3 A global parametrix

In the following theorem Lax's construction is generalized to a global construction on the product of an oriented manifold X and a real line.

Theorem Consider, for $t=t_0$, canonical coordinates (y,η) in a neighborhood M of a point $p=(y_0,\eta_0)$ on $L(t_0)$ and assume that $g(y,\eta)$ defined in M is a very regular phase function for $L(t_0)$ there, i.e. that the variables η can be taken as parameters on $L(t_0)$ in M and that

$$(y,g_y=\eta),\ g_\eta=0,$$

parametrizes $L(t_0)$ in M. Let $(t_0,y,\eta) \to (t,x,\xi)$ be the Hamilton flow and let B be the bicharacteristic issuing from p to $q=(t,x_0,\xi_0)$.

Then there is a number $s>0$ and a conical neighborhood N of p such that, if $0<t-t_0<s$,

A) there are conical neighborhoods N_1 and N_2 of

(x_0,η_0) and (x_0,ξ_0)

respectively such that $(x,\eta)\to(y,\eta)$ is a smooth map from N_1 whose image contains N and $(x,\eta)\to(x,\xi)$ is a smooth map from N_2 whose image contains N_1.

B) there is a neighborhood of N of p such that the equations

$$g(y,\eta) = h(t,x,\eta) = f(t,x,\xi)$$

define functions in N_1 and N_2 respectively, h satisfies the Hamilton-Jacobi equation

$$h_t+q(t,x,h_x)=0,\ h(t,x,s_x)=g(y,\eta) \quad \text{when } t=t_0.$$

and f is a very regular phase function for $L(t)$ in N_2.

Corollary If $a(y,\eta)$ is a polyhomogeneous amplitude functions with conically compact support in N and

$$u(x) = \int a(y,\eta) \exp ig(y,\eta)d\eta$$

is a corresponding oscillatory integral, there is a polyhomogeneous amplitude function $a(t,x,\xi)$ with conically compact support in N_2 such that

$$v(t,x) = \int a(t,x,\xi) \exp if(t,x,\xi)\, d\xi$$

solves the Cauchy problem

$$Pv = \text{smooth},\ v-u= \text{smooth when } t=t_0.$$

Note. From this we can recover a local arametrix. In fact, if we take $t_0=0$, $g=y.\eta$, there is a number s such that N is just the product of a neighborhood of 0 in y-space an all of $R^n\setminus 0$. We can take a=1 and we get a parametrix for small t precisely as at the end of chapter 2.

Proof. The point A) is obvious since x=y, $\xi=\eta$ when $t=t_0$. It follows from A) that the functions h and f are well defined. According to the Theorem of Section 4.1, h solves the Hamilton-Jacobi equation and that f is a very regular phase function follows from the Theorem of section 4.3.

Proof of the corollary. According to B) of the theorem and the theorem of Section 3.3, the Cauchy problem

$$Pu = \text{smooth},\ u-v \text{ smooth when } t=t_0.$$

has a solution

$$u(t,x) = \int b(t,x,\eta) \exp ih(t,x,\eta)\, b(t,x,\eta)\, d\eta$$

with $b(t,x,\eta)$ polyhomogeneous. A change of variables $\eta=\eta(t,x,\xi)$ produces the desired result.

In the proof above, only part of the Theorem of Section 3.3 was used. Using its full power we can prove

Theorem Given $\eta_0\neq 0$ and s>0, there is a conical neighborhood N of $(0,\eta_0)$ such that for any polyhomogeneous amplitude $a(y,\eta)$ with conically compact support in N, the equation

$$Pu(t,x) = \text{smooth},\ u-v \text{ smooth for } t=0.$$

where

$$v= \int a(y,\eta) \exp iy.\eta\, d\eta$$

has a unique solution defined when 0<t<s and in a neighborhood of the projection on X x R of the nul bicharacteristic of P issuing from $(0,\eta_0)$.

Proof. By the previous theorem, the desired result holds for some s>0. And it is clear that if N is shrunk, we can hope for a larger s. By

Hormander's propagation of singularities theorem, the wave front set of u tends to the bicharacteristic B issuing from $(0,\eta_0)$ when N shrinks to the ray generated by η_0. Let r be the least upper bound of numbers t for which u exists near the projection of B on X × R when N is shrunk in this way and let q be the point on B corresponding to r. By the last theorem of chapter 4, there are then canonical coordinates x,ξ at q such that the ξ coordinates are parameters on L(r) and a phase function $g(x,\xi)$, image of $y.\eta$ under the Hamilton flow, which is very regular for L(r) at q. By the preceding theorem, there is a neighboorhood M of q and a number b>0 such that any oscillatory integral w with phase g and amplitude having conically compact support in M can be continued to a solution w' of Pw'=smooth, w'-w smooth for t=r defined when r<t<r+b. Carrying this situation back by the Hamilton flow to t-c with c very small, produces a similar situation with a neighborhoood M_c which permits a similar continuation of oscillatory integrals with conically compact supports in M_c from r-c to to beyond r when c is sufficiently small. This contradicts the definition of r and proves the theorem.

We can now prove that the distribution F(t,x) defined by (5.2.3) and with it the fundamental solution E(t,x) of P with its pole at y, has a global parametrix.

Theorem For any s>0 there is a finite number of functions U(t,x) defined when t<s with the following properties

i) PU is a smooth function

ii) F differs from the sum of the U by a smooth function

iii) U vanishes outside the projection on t,x-space of some bicharacteristic tube T where it has the form

$$U(t,x)= \int a(t,x,\zeta) \exp if(t,x,\zeta)\, d\zeta$$

locally in t. Here the t,z=z(x) are coordinates in the projection, (z,ζ) are canonical. The phase funtion $f(t,z,\zeta)$ is very regular for

$L(t)$ and $a(t,z,\zeta)$ is a polyhomogeneous amplitude with conically compact support in T for t=const.

Proof. By the preceding theorem, we can cover the wave front set of F at the point $t=0, x=0$ by a finite number of conical neighborhoods N such that to every N there is a function U with the properties i) and ii) and such that

$$U(0,x) = \int a(x,y,\eta) \exp i(x-y).\eta \, d\eta$$

where a has conically compact support in N. If we choose these a of degree 0 such that their sum is 1 for all η and all x close to y, the sum of the U solves the same Cauchy problem as F modulo smooth functions. Hence ii) follows and this proves the theorem.

Note. The theorem could be expressed otherwise, namely that so that to any point p in the wave front set of $F(t,x)$ (with t as a parameter) there is an oscillatory integral $U(t,x)$ of the form above such that the wave front set of $F-U$ does not contain a conical neighborhood of p.

Parametrix for a dual pseudodifferential operator

We shall now prove that the taking of duals commutes with the construction of a global parametrix. Together with $P=D_t+Q(t,x,D)$ consider the dual operator $P'=D_t+Q'(t,x,D)$ and let $a'(t,x,\zeta)$ be the duals of the amplitudes $a(t,x,\zeta)$ of the previous theorem.

Lemma Let F' be the fundamental solution of P' with pole at y and let $U(t,x)$ be as in the theorem. Then F' is represented by the distributions

$$U'(t,x) = \int a'(t,x,\zeta) \exp -if(t,x,\zeta) d\zeta$$

in the same way as F is represented by the corresponding distributions $U(t,x)$.

Proof. According to Chapter 3, $(PU)'=P'U'$. Further, the map $U\text{->}U'$ commutes with changes of variables and, as it is easy to verify, also with continuation in the t variable by the Hamilton-Jacobi equation, and with the homogeneous map $(x,\eta)\to(x,\xi)$ under A) of the first theorem of this section. Since these are the only operations involved in the construction of a parametrix of F and since the lemma is obvious for t=, it is proved in general.

CHAPTER 6

CHANGES OF VARIABLES AND DUALITY FOR GENERAL OSCILLATORY INTEGRALS

Introduction Consider a general oscillatory integral

$$u(x) = \int a(x,\theta) \exp is(x,\theta)\, d\theta$$

with x in R^n and θ in R^N and s a regular phase function. It may happen that the number N of integration variables can be reduced locally leading to another oscillatory integral with the same wave front set as the original one. The theory of such changes is due to Hörmander (1971). It occupies the the first two sections of this chapter and will be applied to fundamental solutions in the next chapter.

When

$$a(x,\xi) = \Sigma\, a_k(x,\xi)$$

is a polyhomogeneous amplitude in $R^n \times R^n$, let its dual be

$$a'(x,\xi) = \Sigma\, (-1)^k a_k(x,\xi)$$

and let the dual of the oscillatory integral

$$u(x) = \int a(x,\xi) \exp is(x,\xi)\, d\xi$$

be

$$u'(x) = \int a'(x,\xi) \exp -is(x,\xi)\, d\xi.$$

The dual P' of a polyhomogeneous pseudodifferential operator P is defined as the operator belonging to the dual of its symbol. In connection with the various properties of pseudodifferential operators shown in Chapter 3 we have verified that

$$(P^*)'=(P')^*,\quad (PQ)'=P'Q',\quad (Pu)'=P'u'.$$

In the second part of this chapter we shall see how reduction of the number of integration variables affects the duality of general oscillatory integrals defined in the same way.

6.1 Hörmander's equivalence theorem for oscillatory integrals with regular phase functions

Consider variables x and θ in R^n and R^N respectively and phase functions $f(x,\theta)$ homogeneous of degree 1 in θ and defined in some open set N which is conical in the second variable. Such a phase function is said to be regular if the N differentials df_θ are linearly independent when $f_\theta = 0$. Close to such a point and when $f_\theta=0$, the pair (x,f_x) parametrices a homogeneous Lagrangian manifold L. In fact, the dimension of the manifold is n and when $dx=df_x=0$, then $f_{x\theta}d\theta=0$ and $f_{\theta\theta}d\theta=0$ so that $d\theta=0$. Further, on the manifold we have $\theta.f_\theta=f=0$ so that $0=df=f_x dx$ and changing θ to $t\theta$, $t>0$, changes f_x to tf_x so that L is homogeneous.

It follows form these considerations that wave front sets of oscillatory integrals

$$u = \int a(x,\theta) \exp if(x,\theta)\, d\theta \tag{6.1.1}$$

are contained in homogeneous Lagrangian manifolds.

We shall now consider a situation where two phase functions $f(x,\theta)$ and $g(x,\theta)$ define the same Lagrangian at a point p in the sense that $(x,f_x=g_x)$ at the same point x,θ for which $f_\theta=0$ and $g_\theta=0$ are equivalent equations. It involves the Hessians $f_{\theta\theta}$ and $g_{\theta\theta}$ of f and g and the notion of signature of a real symmetric matrix , i.e. a pair of integers of which the first one is the number of positive and the second one the number of negative eigenvalues of the matrix.

Deformation lemma Under the above conditions,

$$f(x,\theta)= g(x,\theta) + g_\theta.b(x,\theta)g_\theta/2 \tag{6.1.2}$$

close to p where $b(x,\theta)$ is a symmetric matrix. When the Hessians $f_{\theta\theta}$ and $g_{\theta\theta}$ of f and g have the same signature at p, there is a continuous function $b(t)=b(t,x,\theta)$ whose values are symmetric matrices such that $b(1,x,\theta)=b(x,\theta)$ and $b(0,x,\theta)=0$ and the signatures of the Hessians of the corresponding $f(t,x,\theta)$ are constant.

Proof. That (2) holds with some b is clear since $d(f-g)=0$ when $g_\theta=0$. It follows from (2) that

$$f_{\theta\theta} = g_{\theta\theta} + g_{\theta\theta}bg_{\theta\theta} = g_{\theta\theta}(I+bg_{\theta\theta}),$$

so that, by hypothesis, the signatures of $g_{\theta\theta}$ and the right side of the formula are the same. It also follows from (2) and $g_\theta=0$ that

$$f_{x\theta} = g_{x\theta}(I+g_{\theta\theta})$$

so that, since the differentials df_θ are linearly independent,

(6.1.3) $\qquad I+bg_{\theta\theta}$ is invertible.

The desired statement then amounts to the following: given a symmetric matrix A, then two other symmetric matrices B_1 and B_2 can be continuously deformed into each other respecting the condition $\det(I+AB)\neq 0$ if and only if $B+BAB$ has the same signature for $B=B_1,B_2$. When det A is not zero, this is evident for then $I+AB$ and $A+BAB$ are singular at the same time and the map $B\to A+ABA$ is bijective for symmetric matrices B. Hence the statement amounts to the known fact that two non-singular symmetric matrices can be deformed into each other through non-singular matrices if and only if their signatures are the same.

When det A vanishes, let P be the projection on the range of A. Then B can be replaced by PBP without changing any of the two conditions so that we are back in the first case. This finishes the proof if we observe that if a deformation $t\to b(t)$ works at p, it works also in a conical neighborhood of p.

Next, we shall prove part of Hörmander's equivalence theorem for oscillatory integrals (1971, Theorem 3.2.1).

Theorem Let f and g be two regular phase functions defined in a conical neighborhood of a point $p=(x_0,\theta_0)$ and assume that $f_\theta=0$ and $g_\theta=0$ define the same Lagrangian manifold L close to p and that the Hessians of f and g at p have the same signatures. Then there is a bijection

$\theta \to h(x,\theta)$ defined close to p such that $f(x,\theta)=g(x,h(x,\theta))$.

Proof. We have the formula (2) and, by Taylor's formula,

(6.1.4) $g(x,\theta')=g(x,\theta)+(\theta'-\theta).g_\theta+(\theta'-\theta).G_{\theta\theta}(x,\theta',\theta)(\theta'-\theta)$,

where $G_{\theta\theta}$ is a symmetric matrix of homogeneity -1 in θ,θ'. We shall try to determine θ' so that $f(x,\theta)=g(x,\theta')$ by putting

$$\theta' = \theta + w(x,\theta)g_\theta$$

with some symmetric w to be determined. Then, combining (2) and (4) we get

$$z.wz + wz.G(x,\theta,\theta+wz)wz = z.bz/2$$

where $z=g_\theta$. This holds for all z if

$$w+ wG(x,\theta,\theta+wg_\theta)w=b/2$$

which is a complicated equation for w. However, if t is small, b=b(t) is small and there is a unique solution w=w(t) and a corresponding f=f(t) connected to g by a change of the θ variable of the form desired. After this, we can let f(t) play the part of g and proceed another step. We could also have started with g replaced by any f(t) and obtained a change of variables connecting f(t) with some f(s) with s>t. Hence a classical covering argument finishes the proof.

6.2 Reduction of the number of integration variables

Suppose that f is a regular phase function defined in a conical neighborhood N of a point $p=(x_0,\theta_0)$ and suppose that $a(x,\theta)$ is a polyhomogeneous amplitude with conically compact support in N. Then the oscillatory integral

(6.2.1) $$u(x) = \int a(x,\theta) \exp if(x,\theta)\, d\theta$$

is well defined but it is sometimes possible to express u locally as an oscillatory integral with a smaller number of integration variables plus a smooth function. We shall see that if the hessian $f_{\theta\theta}$ has rank r at p, we can in this way eliminate r variables. When A is a symmetric matrix, we define its sign, sgn A, to be $(-1)^d$ where d is the difference between the the number of positive and the number of negative eigenvalues of A.

Theorem Suppose that there is a division of variables θ', θ'' such that the partial Hessian $Q = f_{\theta''\theta''}$ of f at p has rank $r = \dim \theta''$. Then there is a function $h(x,\theta')$, homogeneous of degree 1 in θ', such that

$$g(x,\theta') = f(x,\theta',h(x,\theta'))$$

is a regular phase function close to p which defines the same Lagrangian L there as f and an amplitude $b(x,\theta')$ such that

$$v(x) = c(Q) \int b(x,\theta') \exp ig(x,\theta') \, d\theta' \tag{6.2.2}$$

differs from u close to p by a smooth function. Here

$$c(Q) = (2\pi)^{r/2} \, |\det Q|^{-1/2} \, i^{\mathrm{sgn}\, Q/2} . \tag{6.2.3}$$

When a has the expansion

$$\Sigma a_j , \quad j=k,k-1,\ldots, \tag{6.2.4}$$

b has an expansion

$$\Sigma b_{r/2+j}, \quad j=k,k-1,\ldots \tag{6.2.5}$$

where the indices indicate degree of homogeneity.

Proof. Since Q has rank r, the equations $f_\theta = 0$ determine θ'' as a function $h(x,\theta')$ of x and θ' close to p. Hence, changing θ'' to $\theta'' + h(x,\theta')$ we may assume that h=0, an operation which preserves the homogeneities in (4). Let P be any quadratic form in the variables θ'' which has the same signature as Q. Then the two phase functions

$$f(x,\theta) \text{ and } f'(x,\theta',\theta'') = f(x,\theta',0) + P(\theta'')/2|\theta'|$$

define the same Lagrangian manifold at p and have the same signature there. Hence, by the previous theorem, they are connected by a transformation of the θ variables close to p. Hence if the support of a is conically close enough to p, there is a polyhomogeneous amplitude $a'(x,\theta)$ such that u(x) differs from

(6.2.6) $$u'(x) = \int a'(x,\theta',\theta'') \exp if'(x,\theta',\theta'')\, d\theta$$

by a smooth function. In order to compute the right side of this formula we need a diversion.

Diversion to the method of stationary phase

Lemma Let y,η be real variables in R^r, Q a nondegenerate quadratic form and $f(y)$ a smooth function with compact support. Then we have the following asymptotic expansion for large $t>0$,

(6.2.7) $$\int f(y) \exp itQ(y)/2\, dy = C(Q) \sum t^{r-|\alpha|/2} c_\alpha D^\alpha f(0)$$

with $C(Q)$ according to (3) and where

$$c_\alpha = c_\alpha(Q) = D_\eta{}^\alpha \exp(-iQ^{-1}(\eta)/2)/\alpha! \quad \text{for } \eta = 0$$

vanishes unless $|\alpha|$ is even.

Proof. The Fourier transform of $\exp iQ(y)/2$ is

$$C(Q) \exp -iQ^{-1}(\eta),$$

which is verified by a diagonalization of A, reducing the formula to the case $r=1$. Hence the left side of (7) is $C(Q)$ times

$$(2\pi)^{-r}\, t^{r/2} \int f_\wedge(\eta) \exp(-iQ^{-1}(\eta)/2t)\, d\eta.$$

Expanding the exponential gives a series of terms

$$c_\alpha \eta^\alpha t^{-|\alpha|/2}$$

from which the desired result follows.

Return to the proof of the theorem

By the diversion (and if $\det P = \det Q$), the right side of (6) equals

(6.2.8) $$C(Q) \sum |\theta'|^{(r-|\alpha|)/2} c_\alpha(Q) D_\zeta{}^\alpha a'_j(x,\theta',\zeta)$$

where $\zeta=0$ after the operations. Hence, collecting terms with the same homogeneity, we get

(6.2.9) $$b_{r/2+j} = \sum c_\alpha(Q) D_\zeta{}^\alpha a'_q(x,\theta',\zeta) \quad \text{for } j = q-3|\alpha|/2.$$

This proves the theorem.

6.3 Duality and reduction of the number of variables

We shall now consider duality of oscillatory integrals in the more general setting than we have done so far. In view of our results on the reduction of the number of integration variables, it is then convenient to introduce classes of amplitude functions with integral or half-integral homogeneities. More generally, when m is any real number, let S^m be the class of amplitudes $a(x,\theta)$ with expansions of the form

$$a_{m-p}(x,\theta), \quad p=0,1,\dots \quad .$$

where the indices indicate homogeneity in θ. For oscillatory integrals with N variables it turns out to be natural to normalize both the reduction of the number of integration variables by putting

(6.3.1) $I^{\epsilon}(f,a,x) = (2\pi)^{-N/2} \int \Sigma\, e(\epsilon(q-m))a_{m-q-N}(x,\theta) \exp if(x,\theta)\, d\theta$,

where $q=j-N/2$, $j=0,1,2,\dots$, $\epsilon= 1$ or -1 and $e(j)=i^j$. Note that as ϵ changes sign, terms of corresponding homogeneity m-q-N get multiplied by -1 to the power q-m. Hence the duality differs from the one we have considered before only by a normalization and the fact that m may be a half-integer.

If $b(x,\theta')$ is the amplitude computed in the theorem above and if we put

$$a'(x,\theta') = |\det Q|^{-1/2}\, b(x,\theta'),\ f'(x,\theta')=g(x,\theta'),\ N'=\dim \theta',$$

the oscillating integral (1) and its dual can be written as

(6.3.2) $I^{\epsilon}(f',a',x) =$

$(2\pi)^{-N'/2}\int \Sigma\, e(\epsilon(q'-m))a'_{m-q'-N'} \exp i\epsilon f'(x,\theta')\, d\theta'$.

With these notations, (1) and (2) with $\epsilon=1$ give (6.2.1) and (6.2.2). In particular, the difference between (1) amd (2) is a smooth function. We shall be interested in the result of applying our reduction of the the number of variables also to (1) with $\epsilon= -1$. When A is a symmetric matrix, let $r=r(A)$ be the rank of A and $r_+(A)$ and $r_-(A)$ the number of positive and negative eigenvalues of A and d(A) their difference.

Lemma The difference

$$I^-(f,a,x) - (-1)^t I^-(f',a',x)$$

is smooth when $t=r_-(f_{\theta\theta})-r_-(f'_{\theta'\theta'})$, where the Hessians are taken at (x_0,θ_0) and the corresponding point for θ'.

Proof. In our new normalizations, (6.2.9) reads as

$$e(\epsilon(j'-N'/2-m))\, a'_{m-j'-N'/2} = e(\epsilon d(Q)/2)) \text{ times}$$

$$\Sigma\, e(\epsilon(j-N/2-m))\, c_\alpha(\epsilon Q)|\theta'|^{(r-|\alpha|)/2} D_\zeta^\alpha\, a_{m-N/2-j}(x,\theta',\zeta)|\zeta=0$$

for $\epsilon=1$ and $j'=j+3|\alpha|/2$ making the homogeneities equal since $r(Q)=N-N'$. This formula can also be written as

$$e(\epsilon j')\, b_{m-j'-N'/2} =$$

$$=e(\epsilon(d(Q)+r(Q))/2)\ \Sigma\, e(\epsilon j)|\theta'|^{(r-|\alpha|)/2} c_\alpha(\epsilon Q)\ D_\zeta^\alpha\, a_{m-j-N/2}|\zeta=0,$$

with sum over $j'=j+3|\alpha|/2$. When ϵ changes its sign, the left side is multiplied by -1 to the power j and the terms of the sum on the right by -1 to the power $j+|\alpha|/2$ since $c_\alpha(-Q)= (-1)^{|\alpha|/2} c_\alpha(Q)$ and $|\alpha|$ is even. Since $j+|\alpha|/2$ is congruent to j' mod 2, the change of sign in the formula is determined by the change of sign of the factor in front of the sum on the right. This change is -1 to the power $r_-(Q)$, which in view of the way that Q was obtained, is the same as the number t of the lemma. This finishes the proof.

Paired oscillatory integrals

Let $I^\epsilon(f,a,x)$ be dual oscillatory integrals. The sums

$$J^\epsilon(f,a,x) = I^+(f,a,x) + \epsilon I^-(f,a,x)$$

are said to be paired. It is obvious that the pairing survives any simultaneus change of variables of integration which preserves homogeneity. By our lemma above, a simultaneous reduction of the number of variables in the two terms produces a change of ϵ to $(-1)^t\epsilon$ where t is the number of the lemma. Hence the pairing appears as a very stable operation, unaffected by changes of variables and affected only by a sign when the number of integration variables are changed.

As an example we shall consider fundamental solutions of two dual first order hyperbolic pseudodifferential operators P and P'. If

$$U(t,x) = \int a(x,\xi) \exp if(t,x,\xi)\, d\xi$$

is the local expression for a fundamental solution E of $P = D_t+Q(t,x,D)$ modulo smooth functions, it was proved at the end of the preceding chapter that the corresponding expression for the fundamental solution E' of $P'=D_t+Q'(t,x,D)$ is

$$U'(t,x)= \int a'(x,\xi) \exp -if(t,x,\xi)\, d\xi .$$

Here $a' = \Sigma\ (-1)^k a_k$,k=0,-1,... when $a=\Sigma\ a_k$ with a_k homogeneous of degree k and analogously for P and P'. If U is denoted by $I^\epsilon(f,a,x)$ with $\epsilon=1$, the sum U+U' corresponds in our new normalization to

$$J^\epsilon(f,a,x)$$

where $\epsilon = (-1)^n$. Using the lemma above we can trace its behaviour under changes of the number of variables. The pairings which occur in Lax's construction of a parametrix for the fundamental solution of a strongly hyperbolic differential operators (see the end of Chapter 3) can also be expressed in terms of the new normalization. If f_k, a_k for k=1,2,...,m are the phase functions and amplitudes that occur in the parametrix, it takes the form

$$(J^\epsilon(f_1,a_1,x)+\ldots+J^\epsilon\ (f_m,a_m,x))/2$$

where, again, ϵ is -1 raised to the power n which here is the number of space variables. This value of ϵ accounts for the very different behavior of the singularities of fundamental solution of hyperbolic differential operators in even and odd dimension.

The coming chapter is devoted to the analysis of singularities of paired oscillatory integrals.

Note. The material of this chapter is taken from Hörmander 1971, its application to dual and paired integrals from Gårding 1977. The next chapter gives an expanded version of that paper.

CHAPTER 7

SHARP AND DIFFUSE FRONTS OF PAIRED OSCILLATORY INTEGRALS

Introduction In Chapter 5 we have constructed a global parametrix for the fundamental solution of a hyperbolic first order pseudodifferential operator and with it a gobal parametrix for a strongly hyperbolic differential operator. Locally, these parametrices are oscillating integrals with very regular phase functions. Hence it should be possible to analyze explicitly how they behave near any singular point. As said in the historical introduction, such questions were raised long ago. The first observation came with Poisson's formula for the forward fundamental solution of the wave equation. Its support is the light cone which means that the singularity can not be felt outside the light cone by the behavior of the fundamental solution there. This phenomenon occurs in a weaker form for the fundamental solution of the wave equation in four variables with variable coefficients, constructed by Hadamard. Here the light cone is no longer straight and it is only the singular support of the fundamental solution. But the fundamental solution has smooth extensions over the light cone from either side, by zero from the outside and by something else from the inside. This phenomenon, called a sharp front, reoccurs for an even number of (space + time) variables and does not occur when the number of variables is odd. Further, since Hadamard's construction is only local, these assertions have been proved for small time only.

To catalogue all singularities of oscillatory integrals with regular phase functions is probably impossible. The aim of this chapter is only to give criteria for sharp fronts. They seem to be restricted to paired oscillatory integrals (this may even be proved). The main criterion is a local version of the Petrovsky topological criterion for lacunas which extends to fundamental solutions of hyperbolic differential operators with variable coefficients and to paired oscillatory integrals. Precisely as in the constant coefficient case (see Chapter 1) it applies after a radial integration in the oscillatory integral.

7.1 A family of distributions in one variable

In our radial integrations to come we shall meet distributions in one variable x defined by oscillating integrals of the form

(7.1.1) $\int_0^\infty i^{\epsilon s} \exp \epsilon i x r \, r^{-s-1} f(r)dr$

where ϵ is 1 or -1, x is real and f(r) vanishes for small r and is constant for large r. To this end, it is convenient to consider the following standard integral met with in Section 1.5 in connection with the Herglotz-Petrovsky formula,

(7.1.2) $H(s,z) = \int_0^\infty \exp -rz \, r^{-s-1} \, dr = \Gamma(-s) \, z^s$

where s is real <0 and Re z >0. As explained there, H(s,z) extends to a meromorphic function of s with poles at s=p=0,1,.. and to an analytic function of z in the complex plane cut along the negative axis. For s=p, we define H(p,z) by the formula (1.5.3). Let us note that

(7.1.3) $dH(s,z)/dz = -H(s-1,z)$

for all values of s.

Lemma When f(r) is a smooth function which vanishes for small r and equals 1 for large r, the difference

(7.1.4) $H(s,z) - \int_0^\infty \exp -rz \, f(r) r^{-s-1} dr$

is an entire analytic function of z.

Proof. Since the integral of (4) is an entire function of z when f has compact support in $0<r<\infty$, it suffices to take f(r)=1 when r>1 and zero otherwise. Assume first that s is not an integer p=0,1,2,... . Then the difference above equals the integral

$\int_0^1 \exp -rz \, r^{-s-1} ds$

continued analytically in s from Re $s<0$. This can be done by an expansion of the exponential. The analytical continuation equals

$$\Sigma\ (-z)^{k}/k!(k-s)\quad,\ k=0,1,2,\ldots$$

and is in fact entire analytic in z. Hence the lemma is proved in this case. Next, let $s=p$ be an integer ≥ 0. It suffices to prove that

$$\int_{1}^{\infty} \exp -rz\ r^{-p-1}dr + (-z)^{p}\log z/p!$$

has an analytic extension to $z=0$. When $z>0$ is real, a change of variables in the integral in this formula shows that it equals

$$z^{p} \int_{z}^{\infty} \exp -r\ r^{-p-1}dr$$

and the equality holds by analytical continuation for all $z\neq 0$. Now our last integral differs from

$$z^{p}\int_{z}^{1} \exp -r\ r^{-p-1}dr$$

by a constant times z^{p}. Expanding the exponential in this last integral and doing the integrations produces the convergent sum

$$\Sigma\ (-1)^{k}(1-z^{k-p})/(k-p)k!\quad,\ k=0,1,\ldots\quad,$$

where the term with $k=p$ has to be interpreted as $(-1)^{p}(1-\log z)/p!$. whose singularity is precisely that of $H(p,z)$. This proves the lemma.

We can now compute the integral (1) noting that $(i\epsilon)^{s}=i^{\epsilon s}$.

Theorem The integral (1), i.e.

$$i^{\epsilon s} \int_{0}^{\infty} \exp \epsilon ixr\ \ r^{-s-1} f(r)\ dr$$

where f is a smooth function vanishing close to the origin and constant for large r, equals

$$H(s,x+i\epsilon 0)\ f(\infty).$$

modulo an entire analytic function of x.

Proof. We may assume that $f(\infty)$ equals 1. When Re $i\epsilon x<0$, i.e. Im $\epsilon x>0$, the integral converges and, by the previous lemma equals

$$(i\epsilon)^{s}\ H(s,-i\epsilon x)=(i\epsilon)^{s}(-i\epsilon x)^{s}\Gamma(-s)=H(s,x)$$

modulo an entire function of x. When s is not an integer $p=0,1,2,\ldots$, this equals $H(s,x)$ so that the theorem is proved in that case by analytic continuation. When $s=p$ the expression above equals $H(p,x)$

modulo a polynomial. This proves the theorem.

The distributions $H(s,x+i\epsilon 0)$ and their sums and differences

It is important to have a clear view of the behavior of the distributions $H(s,x+i\epsilon 0)$. They are boundary values of functions of a complex variable z which are analytic in the complex plane cut along the negative real axis. Their singular support is the origin and they are equal when $x>0$. Since

$$(iz)^s = \int_0^\infty \exp -izr \; r^{-s-1} dr,$$

when $\operatorname{Im} z<0$, the fiber over zero in the wave front set of $H(s,x+i\epsilon 0)$ is the positive real axis times ϵ. Since $H(0,z) = \log 1/z = \log 1/|z| - i \arg z$, we have

$$H(0,x+i\epsilon 0) = \log 1/|x| - \epsilon i\pi h(-x)$$

where h is Heaviside's function, i.e. the characteristic function of the positive axis. Hence, in view of (3),

$$H(-1,x+i\epsilon 0) = \mathrm{Pv}\ 1/x + i\pi\epsilon h(-x).$$

For integral s, the distributions $H(s,x)$ are just the successive derivatives and integrals of $H(0,x+i\epsilon 0)$. For $s=1/2$,

$$H(1/2,x+i\epsilon 0) = -2\pi^{1/}(x^{1/2}h(x) + i\epsilon(-x)^{1/2} h(-x))$$

and for general half-integral s they are obtained by integration and differentiation of this distribution.

In connection with paired oscillatory integrals we use the following definition.

Definition The distributions $H^\epsilon(s,x)$, $\epsilon=1$ or -1, are defined as

$$H(s,x+i0)+\epsilon H(s,x-i0).$$

When $\epsilon=-1$, these distributions vanish on the positive axis. They are sharp and diffuse, (i.e. non-sharp), from negative and positive side of the origin according to the following table

	s integer	s half-integer
$\epsilon=1$	d-d	s-d

$\varepsilon=-1$ s-s d-s

7.2 Polar coordinates in paired oscillatory integrals

Consider dual oscillatory integrals given at the end of chapter 6,

$I^{\varepsilon}(f,a,x) = (2\pi)^{-N/2} \int \Sigma \exp if(x,\theta)\ e(\varepsilon q)a_{-q-N}(x,\theta)d\theta$

where $e(q)=i^{q}$, $q=j-N/2$ and $j=p,p+1,\ldots$ for some integer p. We shall introduce polar coordinates with respect to a positive smooth function $c=c(\theta)$, homogeneous of degree 1. Then, replacing θ by $r\theta$ where $c(\theta)=1$, the preceding theorem shows that each term in the sum above has the form

$\int H(q,f(x,\theta)+i\varepsilon 0)a_{-q-N}(x,\theta)\ \omega(\theta)$ + a smooth function

if we disregard the powers of π involved. Here

$$\omega(\theta) = \theta_1 d\theta_2\ldots d\theta_N - \theta_2 d\theta_1\ldots \quad >0 \text{ on } c(\theta)=1.$$

In fact, the radial integration produces the differential $r^{p-q-1}dr$.

Since we are only interested in singularities and the singularities of the singular factor under the sign of integration decrease one step when q increases by an integer, we shall use

$\Sigma\ H(q,f(x,\theta)+i\varepsilon 0))a_{-q-N}(x,\theta)$

as an asymptotic sum for singularities in the sequel. This means that the singularities of the paired oscillatory integral

(7.2.1) $J^{\varepsilon}(f,a,x) = I^{+}(f,a,x) + \varepsilon I^{-}(f,a,x)$

are represented by the asymptotic sum

(7.2.2) $\int \Sigma\ H^{\varepsilon}(q,f(x,\theta)\ a_{-q-N}(x,\theta)w(\theta)$

where $q=j-N/2$, $j=p,p+1,\ldots$.

To the sum above one can apply reduction of the number of integration variables according to the following

Reduction theorem Suppose that in the sum above the Hessian of f at a point (x_0, θ_0) has rank r and r_- negativ eigenvalues. Then it differs by a zero expansion from an expansion of the form

$$\int \Sigma\, H^{\epsilon'}(q', f'(x,\theta')b_{-q'-N'}(x,\theta')w(\theta')$$

where $N'=\dim \theta'=N-r$, $q'=j'-N'/2$, $j'=p,p+1,\ldots$ and ϵ, ϵ' differ by a factor of -1 raised to the power r_-.

Proof. This result differs from the reduction theorem of section 6.3 only in the notations.

Example Global parametrix of the fundamental solution of a strongly hyperbolic operator

We can now state the properties of Lax's local parametrix of the solution $F(t,x)$ of the Cauchy problem of section 3.3. For simplicity, t stands for x_0, x for the other n variables and ξ for the dual variables.

Theorem There are asymptotic expansions for small t,

$$J_k(t,x) = \int \Sigma\, a_{k,1-m-j}(t,x,\xi)\; H^{\epsilon}(j+m-1-n.f_k(t,x,\xi))\; w(\xi)$$

with $\epsilon=(-1)^n$ and $j=0,1,2,\ldots$ whose sum divided by 2 represents the singularities of F.

Proof. By Lax's construction, the parametrix is a sum for $k=1,\ldots,m$ of terms

$$F_k = \int \Sigma\, b_{kj}(t,x,\ )\; \mathrm{exp}\, i f_k(t,x,\xi)d\xi\ ,$$

where $j=1-m,-m,\ldots$. Under the pairing $k\to k'$ we have

(7.2.3) $$b_{k'j} = (-1)^j b_{kj}.$$

If we rewrite F_k as

$$\int \Sigma\, (b_{kj}e(-j-n)\; e(j+n)\; \mathrm{exp}\, i f_k(t,x,\xi)\; d\xi\ ,$$

the singularity expansion of F_k reads

$$\int \Sigma\, (b_{kj}e(-j-n))\; H(j+n, f(x,\xi+i0))\; w(\xi)$$

Changing k to k' we have to change $e(j+n)$ to $e(-j-n)$ and $+i0$ to $-i0$. In view of (3) this means that we get the same expression as before with a factor $(-1)^n$ and $+i0$ changed to $-i0$. This proves the theorem if we change the summation index from j to $1-m-j$.

Pairing and the global parametrix

Starting with the form given above, the construction of a global parametrix for our strongly hyperbolic operator proceeds as for a first order hyperbolic pseudodifferential operator by partitions of unity in ξ-space, canonical changes of variables and extension to larger and larger t along the bicharacteristics by the Hamilton-Jacobi equation. In fact, this construction applies when the parametrix is written as a sum of oscillatory integrals and we have seen that it commutes with the rewriting in terms of paired oscillatory integrals. Hence it applies also to a parametrix in the form given above. The advantage of this form is that, knowing the Hessian of the phase function of any of its terms, we know how to apply reduction of variables till we get a phase functions whose Hessian vanishes at the point we are interested in. Below, there are some simple examples of this analysis. To have more complicated examples, we need to review the theory of almost analytic extensions in the next section.

Examples

Let us consider an oscillatory integral (7.2.1) written in polar coordinates and suppose that the Hessian of $f(x,\theta)$ has corank 1, i.e. rank $N-1$ in a conical neighborhood P of a point $p=(x_0,\theta_0)$ and that the supports of the amplitudes are conically compact subsets of P. Then, by the reduction theorem, the singularities of the integral have an expansion of the form

$$\Sigma\, H^{\epsilon}(j-1/2, f'(x,\theta'))b_{-j-1/2}(x,\theta') \quad , \; c(\theta')=1,\; j=p,p+1,\ldots \quad .$$

with no integration. Applying this to the fundamental solution of a strongly hyperbolic differential operator, we get expansions where the distributions

$$H^{\epsilon}(j+m-1-n/2, f'(t,x,\theta')),\quad j=0,1,2,\ldots$$

are multiplied by smooth functions. The sign ϵ is -1 to the power n-1 plus the number of negative eigenvalues of the Hessian involved. At the outer sheet of the singular support and for small t, there are no negative eigenvalues and the fronts have the type s-s when n is even and d-s when n is odd where the last place indicates the outer front, known to be sharp. Inside the outer sheet and for large t, the nature of singularities depends essentially on the number of negative eigenvalues of the Hessians.

7.3 Almost analytic extensions

Let f(x) be a smooth function and consider

$$f(x+iy) = \Sigma\, f^{(k)}(x)(iy)^k/k! \qquad (7.3.1)$$

as an asymptotic expansion in y. As such it satisfies the equation

$$(d/d\bar{z})f(x+iy)=0,\quad d/d\bar{z}= d/dx +id/dy,$$

for applying d/dx we get the series

$$f'(x) +iyf''(x) +\ldots$$

and applying id/dy we get the series

$$-f'(x) -iyf''(x) +\ldots\ .$$

By a simple extension of a result by Borel (or by a theorem by Whitney), the Taylor series (1) for a smooth function can be provided with factors depending on x with the property of rendering the series convergent and having the same derivatives at the same points ot the real axis as the function f. Hence there are smooth functions f(x+iy), called almost analytic extensions of f for which $(d/d\bar{z})f(x+iy)$ vanishes of infinite order when y=0..

In the general case, for smooth real functions f(x) of several variables $x_1,\ldots,x_n$, we define an almost analytic extension of f to be a function f(x+iy) for which f(x-iy) is the complex conjugate of f(x+iy) and

$$d/d\bar{z}_k f(x+iy)$$

vanish of infinite order for for $y=0$ and $k=1,\ldots,n$. Since Borel's theorem extends to functions of several variables, almost analytic extensions always exist. Denoting almost analytic extensions of f by $AE(f)$, it is clear that

$$AE(f+g)= AE(f)+AE(g),\quad AE(fg)=AE(f)AE(g)$$

and that $AE(1/f) = 1/AE(f)$ exists where f does not vanish. It is also clear that. given f, we can choose $g=AE(f)$ so that g vanishes in an arbitrarily small neighborhood of R^n and $g(x+iy)$ vanishes for x outside the support of f.

7.4 Singularities of paired oscillatory integrals with Hessians of corank 2

Consider a term of the expansion (7.2.2) of a paired oscillatory integral (7.2.1),

(7.4.1) $$\int H^\epsilon(q,f(x,\theta))a(x,\theta)\omega(\theta),\quad q=j-N/2.$$

Let $p=(x_0,\theta_0)$ be a point where the Hessian of f has corank 2. If $a(x,\theta)$ has its support close to p, the reduction theorem says that we can rewrite the integral to a similar one with $\dim\theta=2$, or if we take the polar form into account, $\dim\theta=1$, and with another ϵ and another q. We let (7.4.1) represent this new situation and choose coordinates so that $\omega(\theta)=d\theta$ and $a(x,\theta)$ has compact support on the real line.

By assumption, f is a regular phase function so that f_x is not zero. Let L be the Lagrangian manifold given by f and let X be its projection on x-space. Then X contains x_0 and outside X, f_θ does not vanish so that $f(x)$, defined as the set of θ for which $f(x,\theta)=0$, is a manifold of codimension 1. In our case, $f(x)$ is just a finite collection of points, simple zeros of $f(x,\theta)$, some of which come together as x approaches x_0. Let Y be a component of the complement of X from which x_0 can be reached.

To see what happens when x approaches x_0 in Y, we shall use almost analytic extensions in θ of our functions f, denoted by the same letters. Let M be a narrow strip around the real axis R in the complex plane. Then the differential

(7.4.2) $F(x,\theta) = H(q,f(x,\theta))a(x,\theta)d\theta$

and its differential

$$dF(x,\theta) = (H(q,f(x,\theta)(\partial a/\partial\bar{\theta})(x,\theta) + H'(q,f(x,\theta))(\partial f/\partial\bar{\theta})(x,\theta))d\theta d\bar{\theta}$$

are well defined locally in M as long as Im $f(x,\theta)$ is not zero. When the functions involved are analytic, the differential is zero. In the general case it vanishes of infinite order when Im $\theta=0$, a property which will be a substitute for analyticity. In the sequel, f(x) stands for the zero set of f in M, Re f(x) for its real part and Im f(x) for its imaginary part. The points of Im f(x) then occur in conjugate pairs.

Our program is to rewrite the functions

(7.4.3) $\int H(q,f(x,\theta)+i\epsilon 0)a(x,\theta)d\theta$

as integrals in the complex plane modulo smooth functions. For this we shall use Stoke's formula applied to dF and F and certain regions and their boundaries. The variable t in the definition that follows doubles for a real θ.

Definition For $0<s<1$, let

(7.4.4) $x,t \to v(x,t,s)$

be smooth curves with Im v very small chosen to have the following properties

(i) $t=0 \Rightarrow v=t$

(6.4.5) (ii) $t=0,\ f(x,t)=0 \Rightarrow \text{Im } f_t v_s > 0,$

(iii) $s>\theta \Rightarrow f(x,t) \neq 0.$

That such functions exist is clear. We only have to choose $v=\theta$ outside neighborhoods of points where f vanishes and in these neighborhoods Im v small with the property that (ii) holds. In fact, close to these points,

$f(x,v) = f(x, \mathrm{Re}\, v) + i\, \mathrm{Im}\, v\, f_t(x, \mathrm{Re}\, v) + O((\mathrm{Im}\, v)^2)$.

Lemma Let $c(x)$ be the chain

$$t,s \to v(x,t,s), \quad 0<s<1.$$

oriented by $dtds>0$ and $b(x)$ its boundary at $s=1$ oriented by $dt>0$. Then

(7.4.6) $\int H(q,f(x,\theta)+i0)a(x,\theta)d\theta = \int_{b(x)} F(x,\theta) - \int_{c(x)} dF(x,\theta)$.

Note. The chain $c(x)$ and its boundary $b(x)$ are shown in the figure below, where the zero set of $f(x,\theta)$, θ real is denoted by $f(x)$. When q is a half-integer, $c(x)$ depends on the choice of the square root at one point.

Figure 1 Crosses denote $f(x)$

Proof. By Stokes' formula, it suffices to prove that the left side is the limit of F integrated over the curve $t \to \theta = v(x,t,s)$ as s tends to zero. In other words, if we put

$f(x,v(x,t,s) = f(x,t) + u(x,t,s)$, $v(x,t,s) = t + w(x,t,s)$,

we have to verify that

$$\int H(q,f(x,t)+u(s,t,x))a(x,t+w(x,t,s)\ d(t+w(x,t,s)$$

tends to the left side of (6) as s tends to zero. But this is clear since $u(x,t,s)$ and $w(x,t,s)$ tend to zero with all derivatives when s tends to zero and, in addition, $\mathrm{Im}\, u(x,t,s)>0$. The proof is finished.

Since $dF(x,\theta)$ vanishes of infinite order when $\mathrm{Im}\, \theta$ tends to zero it is obvious that the last term on the left of (6) is a smooth function of x across x_0. Hence the lemma proves that the left side and the first term on the right of (6) differ by a smooth function at x_o.

We are now ready to investigate the singularities of the integrals (1). In doing this we have to distinguish between integral and half-integral q.

Integral q

Theorem. As x tends to x_o in Y, the integral

(7.4.7) $\int H^{\epsilon}(q,f(x,\theta))\ a(x,\theta)d\theta$

has a sharp front at x_o if

i) $\epsilon=1$ and no complex zeros of $f(x,\theta)$ appear

ii) $\epsilon=-1$ no points of Re f(x) come together or meet incoming parts of Im f(x).

Note. The real points of f(x) may come together under i) and the complex points may come together under ii) but they are not allowed to converge to the real points of $f(x_o)$. For general amplitude functions a, the conditions for sharp fronts are also necessary. This can be proved by simple calculations for analytic data. The same remark applies to the next theorem.

Note. Since the theorem is the same for all integral q, it applies to all terms of (7.2.2) when N is even. The corresponding remark applies to the next theorem which concerns half-integral q.

Note. With slightly reformulated conditions, this theorem and the following one hold also for paired oscillatory integrals with dim θ >1. See the end of section 7.5.

Proof. According to the lemma, (7) differs by a smooth function from the integral of $F(x,\theta)$ over

$$b(x)\ +\ \epsilon\bar{b}(x).$$

Hence the theorem follows from the figures below.

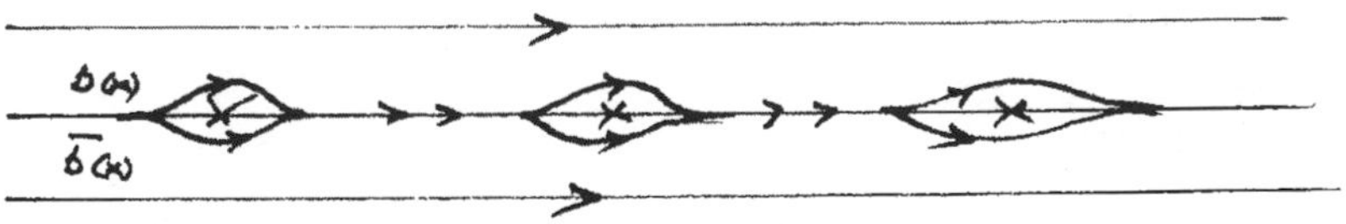

q integer, $\epsilon=1 \Rightarrow b(x)+\bar{b}(x)$ is homologous in $M\setminus f(x)$ to 2 Re M detached from Re f(x)

q integer, $\epsilon=-1 \Rightarrow b(x)-\bar{b}(x)$ is homologous in $M\setminus f(x)$ to loops around Re f(x) with alternating orientations

Half-integral q

We have to modify our former machinery by introducing a two-sheeted cover $M^{1/2}$ of M ramified around Re f(x). Its points are pairs (θ,θ') where

$$\theta'^2 = f(x,\theta).$$

Definition Let $f_\epsilon(x)$ denote the part of the real line where $\epsilon f(x,\theta)>0$ and let c(x) and its conjugate $\bar{c}(x)$ be the same chains as before. Let $c_+(x)$ and $c_-(x)$ be chains obtained by lifting c_* to $M^{1/2}$. Since every half-turn of the cycle b(x) carries (θ,θ') from one sheet to the other, the two chains obtained in this way belong to different sheets of $M^{1/2}$. When M is connected, the construction is unambiguous apart from the choice of θ' at one point. The construction means that the function H(q,f) has alternating signs on the two cycles $b_+(x)$ and $b_-(x)$ between the real zeros of $f(x,\theta)$ and that if we choose H>0 on $b_+(x)$ over an interval of $f_+(x)$, H will be <0 on $b_+(x)$ in the next interval of $f_+(x)$. Hence, if we project the cycles $b_+((x)$ and $b_-(x)$ and their conjugate down to M, we get the following figures

q half-integer, $\epsilon=1$. The cycle $b_+(x)+\bar{b}_+(x)$ projected down to M is homologous in $M\setminus f(x)$ to loops around $f_+(x)$ with alternating orientations

q half-integer, $\epsilon=-1$. The cycle $b_-(x)+\bar{b}_-(x)$ projected down to M is homologous in $M\setminus f(x)$ to loops around $f_-(x)$ with alternating orientations.

The pictures above provide a proof of the following result.

Theorem Wen q is a half-integer, the distributions

$$\int H^\epsilon(q,f(x,\theta))a(x,\theta)d\theta$$

have sharp fronts at x_0 if new zeros appear only outside $f_\epsilon(x)$ and all collapses involve just one component of $f_\epsilon(x)$.

Applications to cusps and swallow's tail

Cusps appear when the phase function has a triple zero. To cover this case, we can assume that there are smooth variables $s=s(x,\theta)$ and $y=y(x)$ such that

$$f(x,\theta) = s^3/3 + y_1 s + y_2$$

close to a point where $f=f_\theta=f_{\theta\theta}=0$ corresponding to $s=y=0$. The curve $f=0$, $f_s=0$ has a cusp dividing the y plane in two regions where

$s \to f(x,\theta)$ has one or three real zeros. The figure below shows the positions of the zeros in the various parts and, when relevant, also the intervals of $f_+(x)$ and $f_-(x)$. According to the rules above, the paired oscillatory integral

$$\int H^{\epsilon}(q, f(x,\theta))a(x,\theta)d\theta$$

has sharp and diffuse fronts according to Figure 4 below.

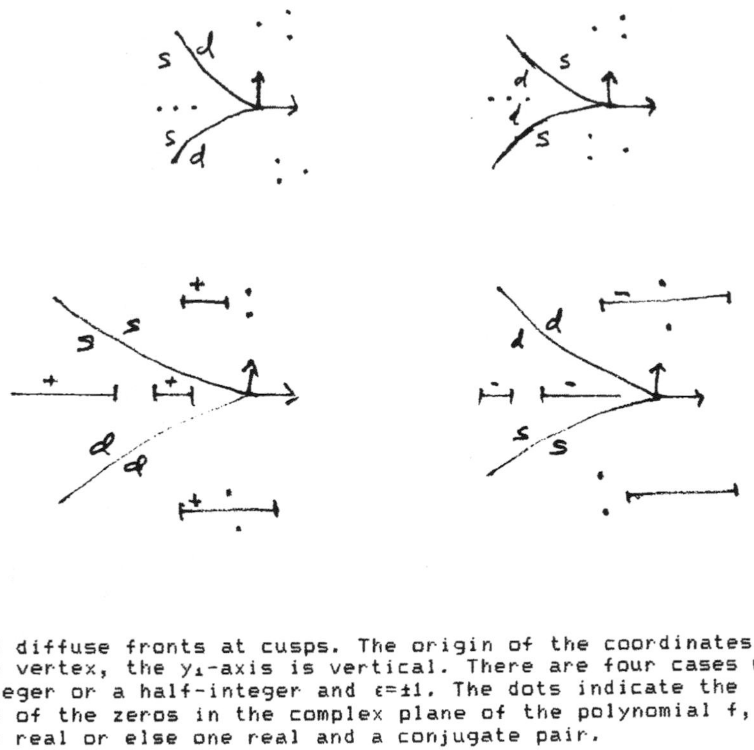

Sharp and diffuse fronts at cusps. The origin of the coordinates y_1, y_2 is at the vertex, the y_1-axis is vertical. There are four cases where q is an integer or a half-integer and $\epsilon=\pm1$. The dots indicate the positions of the zeros in the complex plane of the polynomial f, either all three real or else one real and a conjugate pair.

Next, assume that $f(x,\theta)$ has an isolated singularity as before but one of order 4. Then there are new smooth variables $s=s(x,\theta)$ and $y=y(x)$ such that, after eventually changing f to $-f$,

$$f(x,\theta)= s^4/4 + y_1 s^2/2 + y_2 s + y_3.$$

The surface $f=f_s=0$ is then a swallow's tail whose sections with the plane y_1=const are sketched in the next figure together with the positions of the zeros of f.

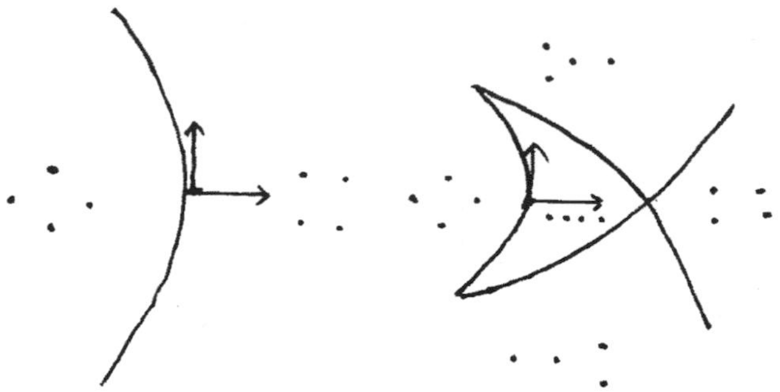

The arrows are at the origin of the coordinates y_2,y_3 with the y_2-axis vertical. The left figure refers to $y_1<0$, the right one to $y_1>0$.

Applying our criteria, we get the figure below to be interpreted as before. To its first column one might add that the corresponding distribution is sharp from inside the tail at $y_1=0$. In case a swallow's tail appears at the outer wave front of the forward fundamental solution of a second order strongly hyperbolic differential operator, the front outside of the tail must have an s on the side facing the complement of the support. This means that only the columns two and four apply, the first one for odd and the other one for even dimension.

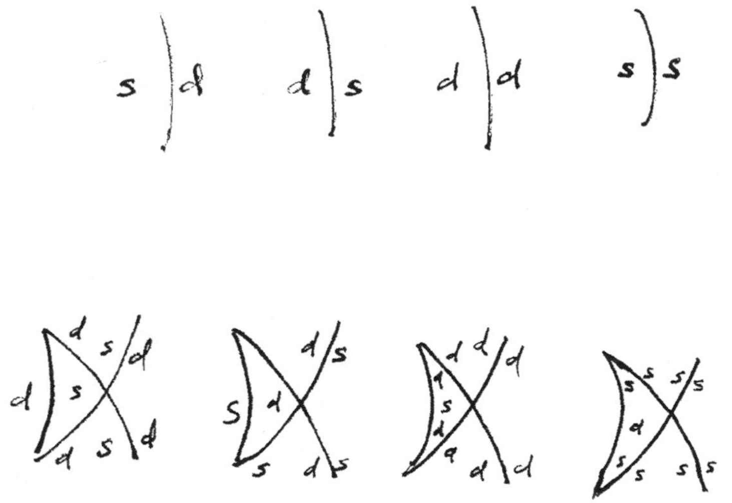

integer,+ integer,- half-integer,+ half-integer,-

Sharp and diffuse fronts at a swallow's tail

7.5 The general case, Petrovsky chains and cycles, the Petrovsky condition

The results of the preceding section extend immediately to arbitrary dim θ. Consider a paired oscillatory integral in polar form

(7.5.1) $$\int H(q,f(x,\theta))a(x,\theta)w(\theta)$$

and let $p=(x_0,\theta_0)$ be a point where f_θ vanishes. Assuming that a has conical support close to some direction, we might as well introduce inhomogeneous coordinates θ with dim θ=N-1 and $w(\theta)=d\theta$. The support of the function $\theta\to a(x,\theta)$ then appears as a bounded open set of R^{N-1}, which, when we keep x close to x_0, is contained in some fixed bounded open set Q. Let X be the projection on x-space of the Lagrangian manifold associated with f, let Y be a component of its complement from which x_0 can be reached. We shall investigate the behavior of the paired oscillatory integral when x approaches x_0 in Y.

When x is outside X, the zero set f(x) of $f(x,\theta)$ is a manifold of codimension 1, separating parts where $f(x,\theta)>0$ and <0. On f(x), the gradient f_θ does not vanish and defines normals to f(x).

Now let $f(x,\theta)$ and $a(x,\theta)$ denote almost analytic extensions in θ of f and a to a strip M around Q in C^{N-1} and let f(x) be the zero set of $f(x,\theta)$ in M. As before, we can construct functions $x,t\to v(x,t,s)$ with the properties (7.4.5) and values in M. This allows us to introduce the differential (N-1)-form

$$F(x,\theta) = H(q,f(x,\theta)d\theta$$

in M\Q. The formula (7.4.6), i.e.

(7.5.2) $$\int H(q,f(x,\theta)+i0))a(x,\theta)d\theta = \int_{b(x)} F(x,\theta) - \int_{c(x)} dF(x,\theta)$$

still holds with c(x) the chain $t,s\to v(t,x,s)$ and b(x) its boundary at s=1, $t\to v(x,t,1)$, still well illustrated by Figure 1. The second term on the right is a smooth function of x up to and including the boundary of Y at x_0 since $dF(x,\theta)$ vanishes of infinite order at Q. Hence, by a theorem of Whitney's it has a smooth extension across X at

x_0.

We can also define the chains $c(q,x,\epsilon)$ and their boundaries $b(q,x,\epsilon)$ precisely as before corresponding to the four cases integral or half-integral q and the sign of ϵ. In this general situations we shall call them Petrovsky chains and cycles since they were first used by Petrovsky. The figures 1 is still relevant but the description of orientations in the figures 2 and 3 becomes more complicated. The sharpness theorem for integral q is true if formulated as follows. The oscillatory integral has a sharp front as x tends to x_0 in Y if

i) $\epsilon=1$ and Im f(x) does not meet Re M

ii) $\epsilon=-1$, no component of Re f(x) collapses, no two components of Re f(x) carrying cycles of opposite orientations meet and no component of Im f(x) meets Re f(x).

The corresponding conditions for half-integral q are:
no two components of $f_\epsilon(x)$ carrying cycles of opposite orientations meet and Im f(x) does not meet $f_\epsilon(x)$.

In all the cases above, sharpness has the same source, namely

The Petrovsky condition There is a (N-2)-cycle B in $M\setminus f(x_0)$ such that, for x in Y and sufficiently close to x_0, the Petrovsky cycles $b(x)=b(q,x,\epsilon)$ are homologous to B in $M\setminus f(x)$.

If C(x) is a chain whose boundary is $B - b(q,x,\epsilon)$, we have

$$\int_{C(x)} dF(x,\theta) = \int_{B(x)} F(x,\theta) - \int_{b(x)} F(x,\theta)$$

When f and a are analytic, dF=0 and the Petrovsky condition implies sharpness. Otherwise, one has to know how C(x) depends on x.

In the cases above, the situation is very simple. Under i) one chooses B as twice M moved away from M into C^{N-1}, under ii) as a (N-1)-cycle enclosing all components of f(x) which come together when x tends to x_o and are enclosed by subcycles of b(q,x,-1) with the same orientation. In the case of half-integral q, B is a cycle chosen in the same way with respect to $f_\epsilon(x)$. In all cases, the dependence on x of C(x) is certainly such that the corresponding integral is a smooth function of x across a neighborhood of x_o. The formal proofs are left to the reader. - It may be possible to prove that the conditions above for sharpness are necessary in many cases.

Note. This chapter is a somewhat expanded version of Gårding 1977.

References

Atiyah M.F., Bott R., Gårding L. Lacunas for hyperbolic differential operators with constant coefficients I,II. Acta Math. 125 (1970) 109-189 and 131 (1973) 145-206.

Born M. and Wolf E. Principles of optics. Fifth. ed. Pergamon Press 1975.

Duistermaat J.J. and Hörmander L. Fourier Integral Operators II. Acta Math. 128 (1972) 183-269.

Gårding L. Sharp fronts of paired oscillatory integrals. Publ. RIMS 12, (1977) suppl. Correction ibid. 13 (1977) 821.

Gelfand I.M. and Shilov G.E. Generalized functions. Vol 1. (Moscow 1958) English translation Academic Press 1964.

Hörmander L. The Analysis of Linear Partial Differential Operators I,II (1983)

-"- -"- III,IV (1985)

-"- Linear Differential Operators. Actes Congr. Int. Math. Nice (1970),121-133.

-"- Fourier Integral Operators I. Acta Math. 127 (1971) 79-183.

Ludwig D. Conical refraction in Crystal Optics and Hydromagnetics. Comm. Pure and Appl. Math. XIV (1961) 113-124.

Maslov V.P. Theory of perturbations and asymptotic methods. Mosc. Gos. Univ. Moscow 1965.

Uhlmann G.A. Light intensity distribution in conical refraction. Comm. Pure App. Math. XXXV (1982) 69-80.

Index

almost analytic extension 111
conical refraction 3,24
crystal optics 3
double refraction 4
Fourier integral operators 44
front, sharp or diffuse 107,115,117,118,120,122
fundamental solution 6
Hadamard 6
Herglotz-Petrovsky formula 25,31
homogeneous hyperbolic 13
Hörmander 16,33,36,55,68,96
hyperbolicity cone 14
intrinsic hyperbolicity 17
microhyperbolic 10
Kovalevskaya 4
Lagrangian planes 77
Lax 7,46,70
localization 21
oscillatory integrals 39
-equivalence of 96
-duality of 101
-paired 102
parametrices 46
-global 89,110
Petrovsky criterion,condition 32,104
polyhomogeneous operators 54
propagation cone 20
propagation of singularities 68
pseudodifferential operators 52
-on manifolds 62
-Cauchy's problem for 85,88
symplectic geometry 75
wave equation 2
wave front set 34
wave front surface 21
very regular phase function 83
Volterra 4
Zeilon 4